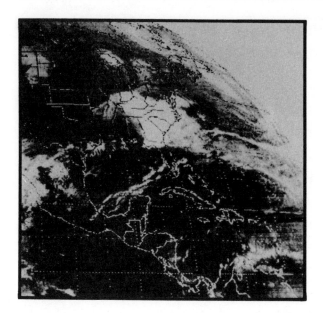

WEATHER SATELLITE HANDBOOK

——— FOURTH EDITION

DR. RALPH E. TAGGART, WB8DQT

A high-resolution segment of a Soviet METEOR visible-light pass in January, 1990. The circular feature near top center is Lake Manicouagan, a pair of semicircular lakes in Quebec. The lakes are ice- and snow-covered in this season, and quite prominent. This feature is thought to represent an ancient meteor impact crater. Immediately below the lake is the Gulf of St. Lawrence with Anicosti island to the right. The image was captured with the scan converter described in Chapters 5 and 6, computer enhanced with a log-black processing curve (Chapter 10) and printed on a Hewlett-Packard DeskJet printer (Chapter 9).

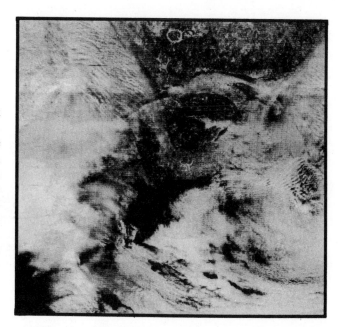

Foreword

Weather . . . This marvelous end result of so many natural forces at work in our global environment is of great importance to us all. At the very least, it can determine our activities for a particular day. At most, it is a force that can sustain—or destroy—life. Because of this, we need to know all we can about the weather. We try to predict it; sometimes, we're successsful. Someday—we dream—we'll even control it.

Before that happens, we've still got a lot to learn about weather. In the effort to increase our knowledge of the forces of weather, one of the steps science has taken is to send satellites into space to give us another viewing platform. These eyes in the sky communicate with us. And if there's one thing radio amateurs know about, it's communications.

Since the launch of Sputnik in October 1957, Amateur Radio operators have been involved in space-related activities. Although amateur-built and operated communications satellites represent the most obvious facet of this involvement, radio amateurs have also been at the forefront in other areas, including the development of ground-station equipment to use the information presented to us by the ever-more-diverse range of weather-satellite signals.

A pioneering effort by Wendell Anderson, K2RNF, described in a November 1965 *QST* article,[1] proved that it was possible to receive and display fascinating pictures of the earth from space using low-cost equipment. Not surprisingly, it has been Amateur Radio operators, more than any other group, who have helped develop the technology to make weather-satellite information accessible with modest ground-station equipment.

The decision of the ARRL to publish this latest edition of the *Weather Satellite Handbook* reflects the role of amateurs in enhancing the understanding of our world through the medium of radio and image communications. The history of Amateur Radio has been a continuing story of the advancement of communications technology. Those involved in making weather-satellite images accessible to the public have been true to this tradition.

David Sumner, K1ZZ
Executive Vice President
June 1990

[1] W. Anderson, "Amateur Reception of Weather Satellite Picture Transmission," *QST*, Nov 1965, pp 11-17.

Preface

The starkly angular satellite, about the size of a telephone booth, moves soundlessly along its orbital track over 800 km above the darkened surface of the earth below. A single wing-like solar panel, coldly bathed in starlight, is useless at the moment: The electronic heartbeat of the machine is now sustained by on-board batteries.

Deep within the spacecraft, an intricate combination of motors, lenses, mirrors and solid-state sensors scans the earth below. Some of the sensors respond to visible light. Right now, they stare blindly at the great black void of the darkened planet beneath. But other sensors—those that respond to heat variations imperceptible to the human eye—are building up an image of shorelines, sea ice, frigid clouds, and occasional patches of open ocean in the arctic night below. Computer circuits process the stream of data from the sensors and route it to several transmitters connected to spidery antennas on the underbelly of the spacecraft.

Within minutes, the horizon ahead and to the left of the spacecraft brightens. The satellite moves steadily toward the light, welcoming the beauty of an orbital sunrise it has witnessed many times before. Now, details of the earth below begin to flood the visible-light sensors. The spacecraft is arcing over extreme northern Canada. Below, Ellesmere Island is wrapped in an arctic twilight and the grandeur of ice-locked Hudson Bay comes into view.

Distorted by the curve of the earth, the Great Lakes seem to roll into view. At that moment, the faint radio signals from the spacecraft activate a receiver in central Michigan. The warbling tone the receiver hears is noisy at first, but strengthens steadily as the satellite rises over the northern horizon. A tape recorder springs to life, ready to preserve a record of the stream of image data transmitted by the unseen traveler. On a television screen, the high-altitude artist begins to paint a picture, line by line. The hypnotic tone of the incoming signal is gradually converted to an image of light and shadow, reproducing the distant observer's view of the sweep of Hudson Bay. As the satellite sweeps southward, the images of the Bay are replaced by a stippled pattern characteristic of snow-covered conifer forests. Soon, the spacecraft is over the Great Lakes and their characteristic form—easily identified in all seasons—is reproduced on the screen. Off to the east, the outline of Cape Cod, followed by Long Island and the coast of New Jersey and Delaware, make their appearance.

During the spacecraft's sweep over the heartland of the country below, the patterns of cloud and snow cover begin to change. By the time the satellite is relaying the image of Florida and the Gulf Coast, every trace of snow is gone; the land below is traced by the patterns of clouds defining a major weather front.

Pushing onward, the spacecraft continues its ceaseless scanning of the earth below. As it passes south of Cuba—with South America looming ahead—the Great Lakes, again distorted by the curvature of the earth, fade into the horizon behind it. Without the space voyager's voice to sustain them, the receiver and tape recorder in Michigan become silent. A computer senses the change. After checking its internal clock, it changes the receiver's operating frequency, then patiently waits. Another visitor is due over the southern horizon in precisely 37 minutes . . . .

You might think the foregoing narrative contains a description of a government installation built to track destructive storms and improve the reliability of weather forecasts. Actually, the system described is in my own basement. It, and thousands of others like it, was built solely to gratify my own interest in watching the ever-changing panorama of the earth as viewed from space. This *Handbook*, now in its fourth edition, is meant to introduce you to that fascinating activity.

The *Weather Satellite Handbook* has changed considerably since it was first published in 1976. These changes have been driven by the increasing sophistication of the satellites themselves, as well as by the steady march of electronics technology that has made it even easier to watch these elusive images. In this edition, the various display projects are entirely digital, reflecting the ease and flexibility provided by modern computer technology. Although the projects themselves are more sophisticated than earlier ones, they are actually simpler to build, and typically cost less than the analog technology that dominated earlier editions of this book.

This hobby is a unique blend of electronics, meteorology, earth science, astronautics, and, now, computer science. It is an activity that has captivated me—to the occasional despair of my family as activities are planned around the orbital timetable of unseen satellites—for over 17 years now. Since it was first published, the *Weather Satellite Handbook* has served to introduce thousands to a fascinating hobby. If this fourth edition does the same for you, it will have fulfilled its purpose.

Mason, Michigan
January, 1990

Lead photo:
Artist's conception of the TIROS spacecraft in orbit, courtesy of the National Oceanic and Atmospheric Administration.

About the Author

Ralph E. Taggart, WB8DQT

Ralph Taggart was born in Charlottesville, Virginia, in 1941. He grew up in a log cabin in the mountains of northern New Jersey, and was licensed as WA2EMC in the late '50s while in high school. He attended Rutgers University and received a BA in Biology in 1963. For two years, Ralph worked as a research assistant at the Boyce Thompson Institute in Yonkers, New York, and spent his spare time working with Lew West (W2PMV) and a small group of northern New Jersey ATV experimenters attempting to establish two-way television communications on the 420-MHz band.

In 1965, Ralph went to Ohio University in Athens, Ohio, to work on a Master's degree. The entire rack of home-brew TV cameras, flying-spot scanners, modulators and transmitters went along for the duration, sparking the first two-way ATV activity in that part of Ohio. He received his MS in Botany in 1967, then journeyed to East Lansing, Michigan, to work on a doctoral program in paleobotany (fossil plants) at Michigan State University in 1967.

Frustrated with the range limitations of conventional television, he built his first slow-scan camera and monitor during that first year at MSU. This equipment provided the means for a number of pioneering activities with SSTV through the Michigan State Amateur Radio Club station, W8SH. Among the more notable achievements of this period was the first transmission of color-SSTV images and the first two-way color SSTV contact. It was during this period that Ralph upgraded to Advanced class and received the call sign WB8DQT.

Ralph received his PhD in 1970, and joined the staff at MSU where he now holds an appointment as full Professor in the Departments of Botany and Plant Pathology and the Department of Geological Sciences. In 1972, he co-authored (with Don Miller, W9NTP) the *Slow Scan Television Handbook*, the first comprehensive introduction to SSTV. About the same time, he became interested in the application of SSTV display techniques to weather-satellite images, leading in 1976 to the publication of the first edition of the *Weather Satellite Handbook*.

Ralph is married, has three daughters, and lives in Mason, Michigan. He is an Elder in the Presbyterian Church, teaches Church School and has served on the local cable-television advisory commission. He is now in his twelfth year on the Mason Board of Education, where he has served as President for the past two years. In addition to his ongoing satellite activities, Ralph is active on UHF ATV and HF SSTV, and enjoys rag-chewing on 80-meter CW. He has also been an avid ultralight pilot for the past nine years, and is just completing a home-built ultralight gyroplane incorporating a number of his own design features.

In addition to his book credits in the area of SSTV and weather satellites, he is the co-author of two general biology textbooks, numerous research papers and monographs and a chapter on ski flying in a book on ultralight flying techniques. Ralph has written a number of magazine articles on ATV, SSTV, weather satellites, and other subjects that have appeared in *QST*, *73 Magazine*, and *Ham Radio Magazine*.

Dedication

If you are about to become committed to pursuing weather satellites, put aside thoughts of hardware and software. Your greatest asset is an understanding spouse who will tolerate the numerous eccentricities that the hobby demands. I am extremely fortunate to have such an individual in my life: my wife, Alison.

In the earliest days of our marriage, Alison tolerated a student apartment crowded to overflowing with Amateur Radio equipment. Even now she faces a basement filled with ever-more-numerous items of equipment—and nothing ever seems to get thrown away. At times, our life seems dominated by the schedules of unseen satellites, phone calls at all hours from half the world away, radio schedules, equipment tests, and even the writing binge necessary to finish this edition. Through all this, she, and my daughters, Jennifer, Heather, and Molly, have been content merely to mutter at activities that would be cause for institutionalization in a less-tolerant household.

Paul Segal (ex-W9EEA), in the *Amateur's Code*, wisely proclaimed that "The Amateur is Balanced." We may try to achieve that standard but, gripped by the fascination of our hobby, we will almost always fail to some extent. I will promise my bonny Alison that, now that the new edition is finished, I will spend more time with the family. I will promise—and inevitably, I will fail. The marvel is that she will not take that failure as seriously as is her right. For that I am profoundly grateful.

Trademarks and Registered Trademarks Used in this Book

Amiga	Commodore Business Machines, Inc
AMSAT	The Radio Amateur Satellite Corp
Heliax	Andrew Corp
Apple II	Apple Computer, Inc
Apple //e	Apple Computer, Inc
Color Computer	Tandy Corporation
Commodore	Commodore Business Machines, Inc
Compuserve	CompuServe Inc
GIF	CompuServe Inc
Hewlett-Packard DeskJet	Hewlett-Packard
IBM	International Business Machines Corp
Kodabrome RC	Eastman Kodak Co
Kodak Co Dektol	Eastman Kodak Co
Kodak Co Microdol X	Eastman Kodak Co
Kodak Co Plus-X	Eastman Kodak Co
Kodak Co Rapid Fixer	Eastman Kodak Co
Macintosh	Apple Computer, Inc
MS	Microsoft Corp
MS-DOS	Microsoft Corp
PC/AT	International Business Machines Corp
PC DOS	International Business Machines Corp
PC/XT	International Business Machines Corp
Pentax	Pentax Corp
Polaroid	Polaroid Corp
QuickBASIC	Microsoft Corp
Radio Shack	Tandy Corporation
Sams Photofact	Howard W. Sams & Co, Inc
Sony	Sony Corporation of America

Contents

Chapter 3: Weather-Satellite Receivers

Chapter 4: Video Formats and Display Systems

Chapter 5: The WSH Microcontroller

Chapter 6: Scan-Converter Display Board

Chapter 7: Scan-Converter and Computer Interfacing

Introduction
Some Notes On Computers
Living With a Computer
Interface Hardware
     A-Bus System
     MetraByte Corporation
Software
     Mode/Function Requests
          Display Functions
          Operating Modes
     Saving a 1700 Image
     Sending an Image to the Scan Converter
     General Notes
     High-Resolution Data
          Standby
          Image Start
          Image Data Stream
Commercial Software
     VGA1700 Program
System Expansion

Chapter 8: Satellite Tracking

Introduction
Orbital-Plotting Board
Satellite Orbits
Predict-Data Format
Predicting Crossings
Where Is the Satellite?
Southern-Hemisphere Satellite Tracking
Longer-Term Predictions
The Concept of Pass Windows
A Simple Predict Program
     Entering the Program Listing
     Language and Computer Compatibility
     Running the Program
     Customizing the Program
     Updating Satellite Data
     Long-term Accuracy
     Other Tracking Programs
Geostationary-Antenna Bearings

Chapter 9: Station Operations

Chapter 10: Advanced Applications

Glossary

Appendix I: Parts and Equipment Suppliers

Appendix II: Scan-Converter Parts List

Appendix III: WSH1700 BASIC Program Listing

Chapter 1

Operational Satellite Systems

INTRODUCTION

There is little doubt that the weather-satellite program of the United States is one of the most tangible benefits of the overall space program. Since the launch of the first operational TIROS satellite in the early '60s, uncounted lives and dollars have been saved as a result of our ability to observe weather phenomena on a global scale. The TIROS satellites of the '60s—primitive by current standards—have been followed by satellites of ever-increasing sophistication. Today, a number of nations are involved in weather observation from space, including the US, the USSR, a consortium of European nations that make up the European Space Research Organization (ESRO), and Japan. The People's Republic of China is expected to launch its first weather satellite sometime in 1990-1991.

Most of these operational satellites fulfill their missions with the transmission of very-high-resolution digital images to specially equipped ground stations. It isn't impossible for amateur stations to receive and display such pictures (Figure 1.1), but it does require a good background in microwave techniques and digital electronics. Such stations are complex and somewhat costly. Eventually, if the past is any guide, the activities of pioneering amateur experimenters will make such installations easier and more economical to assemble.

Fortunately, these same satellites transmit lower-resolution analog images that are specifically designed to be easy to receive and display. These images provide the mainstay for amateur weather-satellite activities. Today's operational satellites fall into two general categories: satellites in low, near-polar orbits, and satellites in geostationary orbits.

POLAR-ORBITING SATELLITES

Current operational polar-orbiting satellites include the TIROS/NOAA series operated by the United States and the Meteor/COSMOS satellites of the Soviet Union. These satellites operate in relatively low orbits (in the case of the US satellites, at altitudes of about 600 miles above the earth) and their orbital tracks are such that they come very close to passing over the poles during each revolution of the earth. Essentially, they pass twice over all parts of the earth every 24 hours, once during the day, and again at night.

US TIROS/NOAA Polar Orbiters

The US TIROS/NOAA satellites are very precisely oriented in space in what are know as *sun-synchronous orbits*. This means that during the course of a year, the relationship of the satellite's orbital track to the position of the sun stays relatively constant. Therefore, the satellite passes overhead at about the same solar time each day. Generally, the Soviet satellites are not in sun-synchronous orbits, so that the time for optimum passes each day changes in the course of a year.

Although the early weather satellites used TV cameras to obtain their pictures, the delicate vidicon camera tubes they employed are easily damaged, and the performance of a given tube deteriorates steadily with time, leading to marginal images. All of today's operational polar orbiters have replaced the vidicon TV pickup tubes with electromechanical systems known as *scanning radiometers*.

A scanning radiometer is basically a system of lenses, a motor-driven mirror system, and one or more solid-state light sensors that serves as the source of the satellite image data. Basically, the scanning radiometer looks at a very narrow line, equivalent to the horizontal line of a TV picture, at right angles to the satellite's orbital track. The equivalent of the vertical scanning in a TV picture is provided by the motion of the satellite along its orbital track. The scanning system operates continuously, so you can receive a picture that spans the full time that the satellite is within range of your station (Figure 1.2).

Although most of the operations of these satellites can be modified by signals from specially equipped ground stations, picture transmission is essentially automatic and continuous. The pictures that we can receive are usually referred to as *Automatic Picture Transmission*, or APT, imagery.

Figure 1.1—A portion of an HRPT image of the central east coast of the United States as received and processed by John Dubois. John has been one of the major forces behind the development of HRPT ground-station equipment that can be duplicated by advanced amateur satellite enthusiasts. It would be possible to write a history of the Civil War using as a background the area encompassed by the southern half of this image. The center is dominated by the great peninsula, containing parts of Delaware and Maryland, bounded to the south by Chesapeake Bay and to the north by Delaware Bay. The west shore of the Chesapeake is dissected by the tidal estuaries of some of the legendary rivers that played a major role in that conflict. To the extreme south is the James River, with the Chickahominy River flowing into it from the north. Further westward is the confluence of the James and the Appomattox Rivers, with the city of Richmond appearing as a small smudge on the James. Immediately north of the James estuary is the York River; north of that is the Rappahannock. The widest of the tidal estuaries, and the next one north, is the Potomac with the city of Washington appearing as a darker smudge where the river abruptly narrows. The city of Baltimore is a slightly larger smudge at the head of the short, but wide, estuary near the northern end of the Chesapeake. The major river flowing into the northern end of the Bay is the Susquehanna. The Delaware flows into the northern end of Delaware Bay, and Philadelphia is clearly visible. In the extreme upper right is the New York metropolitan area, with the wide reach of the lower Hudson River. The many lakes in the Ramapo Mountains area of northern New Jersey and southern New York State are also visible. This particular image was downloaded from the Dallas Remote Imaging Group (DRIG) satellite electronic bulletin board as a Compuserve GIF image, and displayed on an IBM-compatible VGA display system.

Figure 1.2—A NOAA-11 pass covering most of western North America. Mexico, the US Gulf coast, and Baja California, are the most prominent features in the southern half of the image. There is the typical fog bank off the northern California coast and low fog fills the Central Valley. A major cloud system blankets northern Oregon, Washington, and southern British Columbia, including Vancouver Island. This weather system extends inland and covers much of southern Alberta and Saskatchewan. Manitoba is generally clear, and Lakes Winnepeg and Manitoba are clearly visible as is the western tip of Lake Superior to the southeast. In the extreme upper right is a bit of the coastline of western Hudson Bay. This pass was received by manually tracking the satellite, using a small, 5-element beam. Slightly over 13 minutes of the pass were recorded, and about 12 minutes of that are displayed here. Because the scan converter displays slightly over 6 minutes of data, the northern and southern halves of the pass were displayed and photographed separately, then joined to prepare this full-pass mosaic. This pass originated in the south, so the original image would have appeared upside down on the display. I have inverted the image here so that north is at the top, making it a bit easier to orient the major features.

The image that is produced is mostly a function of the type of solid-state image sensor being used. Some sensors respond to various portions of the visible-light spectrum; these create images similar, although not necessarily identical, to those you would obtain in a standard photograph (Figures 1.2 and 1.4). The US TIROS/NOAA satellites have two different visible-light sensors, each with slightly different imaging characteristics. In addition to visible-light detectors, the US satellites also have a number of sensors that respond

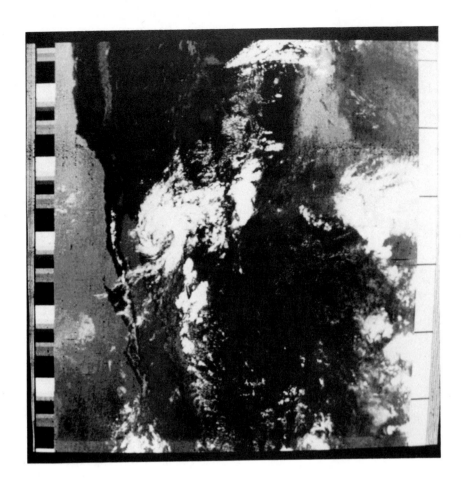

Figure 1.3—The southern half of the same NOAA-11 pass illustrated in Figure 1.2, but showing the IR image data. In the IR format, cold objects (such as high clouds) appear white, while warmer objects are progressively darker. The land masses of Mexico, Baja California, and California are quite dark (hot), contrasting strongly with the cold Pacific Ocean waters to the north. San Francisco Bay is clearly evident on the coast at the upper left and the fog-filled Central Valley shows up well against the warmer surrounding terrain. Note that the offshore fog does not appear in this image. The fog is very low and essentially at sea-surface temperature, thus it does not contrast with the ocean surface at IR wavelengths.

to infrared (IR) or heat radiation. Because each of the IR sensors responds to heat and not light, IR cloud pictures can be obtained at night as easily as during the day (Figure 1.3).

The scanning radiometer of a US satellite is actually obtaining a very-high-resolution image at several IR and visible-light wavelengths simultaneously, but this image is transmitted to earth as a wide-bandwidth digital signal that requires very sophisticated equipment to receive and display. Figure 1.1 shows a sample of a portion of this *High Resolution Picture Transmission* (HRPT) format. The high-resolution imaging system is known as a *multispectral scanning radiometer.* Fortunately, an on-board computer is also sampling the high-resolution picture and transmitting those samples in the relatively simple analog APT format that can be received and displayed with comparatively unsophisticated equipment.

Although the various polar-orbiting satellites are quite different from one another, the basic format of their picture signals have a number of common features that make it easy to design a system for handling all of them. Details of the basic video-modulation format are covered in Chapter 4.

The various satellites differ from one another in the rate at which the video lines are transmitted and the

available image formats (visible/IR). The TIROS/ NOAA APT transmissions use rates of 120 lines per minute (LPM), and each line is made up of both visible and IR image data. The first half of each line represents the IR data scan, while the second half is the visible-light data. If the image is displayed at 120 LPM, the IR and visible-light images appear side by side. Either the IR or visible-light data can be displayed alone by using a 240-LPM display rate and blanking out every other line. This is what has been done to display the NOAA visible-light image in Figure 1.2 and the IR image of Figure 1.3.

At high latitudes, the quality of visible-light imagery varies with the time of day and the season. The requirement for sun-synchronous orbits restricts the TIROS/NOAA satellites to mid-morning and mid-afternoon passes. During the summer months, the illumination angles are excellent. During the winter, sun angles are lower, and one side of the image is brightly illuminated, while the side farthest from the sun (west in the morning, east in the afternoon) is noticeably darker, particularly for early or late passes.

IR pictures (Figure 1.3) are typically disappointing to those expecting the high contrast typical of visible-light data. In the IR format, warm objects are black, and cold objects are white. The ability to differentiate

Figure 1.4—An example of typical Meteor warm-weather imagery. Like the passes illustrated in Figure 1.2 and 1.3, this pass originated in the south and ended in the north. This print was inverted to place north at the top. The distinctive 120-LPM Meteor sync-pulse train thus appears at the extreme right, but it actually marks the start of each Meteor image line. As is typical of Meteor images in the summer, cloud details are excellent, but ground features are essentially nil. Michigan and the Great Lakes are just above and to the right of the center of the image. They are evident because the innumerable late-afternoon cumulus-cloud centers do not develop over the cooler lake waters. This particular image was subjected to a considerable amount of logarithmic image contrast in an attempt to resolve ground detail. The only new feature revealed by this processing is a faint sun-glint off the waters of Lake Michigan. With the lake thus defined, you can see how thunderstorm cell development is inhibited near the lake shore and over the lake itself.

land, water, and cloud features depends on temperature differences. Ample contrast is available at tropical latitudes and desert areas, but contrast is reduced at higher latitudes. Daylight IR typically has greater contrast than night imagery due to an enhanced thermal gradient, and summer images are better than winter ones. Winter night images at high latitudes may appear almost white, with little or no contrast. IR imaging has great utility because it is not tied to local lighting conditions, but specialized techniques for image enhancement are required to make the most of the small variations when ground temperatures range from cool to cold and certain sensors are in use. Most of the year, it is possible to get useful pictures without special processing, but careful adjustment of video gain is required. The reward is the ability to use evening passes to produce useful imagery, and bring out features (such as the Gulf Stream) that are simply not available in visible-light images.

Soviet Polar Orbiters

The workhorse of the Soviet operational satellite system is their Meteor satellite series; these transmit visible-light images at a rate of 120 LPM (Figure 1.4). The Soviet Meteor satellites do not use sun-synchronous orbits, hence, they pass over in daylight at different times during their operational lifetime. If a Meteor satellite passes over in early to mid-morning or mid- to late-afternoon, the imagery has the same unequal lighting characteristic of the TIROS/NOAA pictures during the winter months. The Soviet satellites tend to have a number of operational satellites in orbit at any time, however, and there is usually one with a useful pass near midday, avoiding the winter-lighting problem.

The US TIROS/NOAA satellites have sensors with a very wide dynamic range, making it possible to differentiate land and water features where lighting is adequate. Meteor sensors tend to compress the black end of the gray scale. Clouds are rendered in great detail, but land/water boundaries are almost indistinguishable without extreme video processing (Figure 1.4). During the winter months at high latitudes, snow and ice cover increases the contrast between land and open water and midday Meteor images can provide excellent results.

For many years, Meteor satellites have provided only visible-light imagery and the satellite transmissions would shut off automatically when light levels dropped below certain thresholds. For the past year or so, the Soviets have apparently been experimenting

Figure 1.5—Approximately 12 minutes of a south-to-north pass of Meteor 3.3 illustrating the new Soviet 120-LPM IR format. This pass occurred at about 2350Z on a bitter-cold December evening. Most of the Great Lakes are covered by very cold clouds, although enough of the comparatively warm lake waters are visible to help you orient on the major features. The Carolina coast, Delaware, New Jersey, Long Island, and Cape Cod are warmer than the interior, but still contrast against the warmer ocean water. A major cyclonic storm system covers the Canadian Maritimes, but that is not new to those folks! No image enhancement—other than the pixel complementation discussed in the text—was used for this image. It appears that this format has excellent potential for stations that lack the facility to perform image enhancement on winter IR-image data. The familiar Meteor sync-pulse train is off to the right because the mosaic is inverted to place north at the top.

with a low-resolution IR-imaging system operating at about 20 LPM. The results were far from exciting, but it was encouraging that IR experimentation was under way. Late in 1989, the Soviets launched a new satellite, Meteor 3/3. This satellite has a 120-LPM IR-imaging system that is currently providing excellent night IR coverage (Figure 1.5). One anomaly of this format is that cold areas (such as clouds) reproduce as black, while warm areas are white. Digital pixel complementation was used to convert the original data in

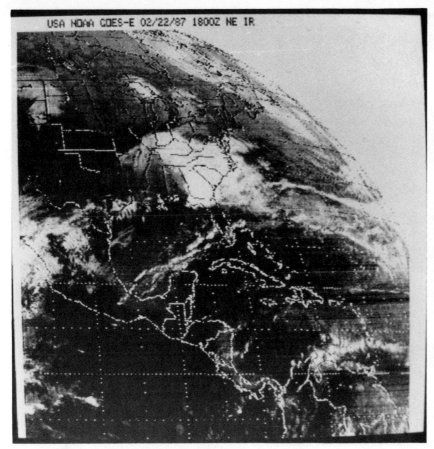

Figure 1.6—A NE IR quadrant, imaged by the GOES East satellite and received via the GOES Central satellite. High, cold clouds are white in the IR format, as is the view of space beyond the limb of the earth. Ground computers insert the political and geographic reference data because some ground features may be obscure at IR wavelengths. This image covers most of North America, all of Central America, and the extreme NW portion of South America. Because the GOES satellites maintain the same position relative to any point on the earth below, a view such as this provides the same coverage in any transmission as long as the satellite sub-point position is maintained. In this sample, a major cloud system blankets the southeastern US, while the Caribbean area is clear. This image was obtained in late February (see the header at the top of the image) and the lower ground temperatures of the northern states and Canada are clearly evident in the lighter tone of these areas.

Figure 1.5 to the more familiar cold = white/warm = black format. (This process is described in greater detail in Chapter 7.) It is too early to tell if this precise format, or some variant of it, will become operational for new Meteor satellites, but given the quality of the imagery, developments will be watched with interest.

The Soviets also operate a higher-resolution satellite system that generates superb images at a rate of 240 LPM (4 lines per second). These pictures—usually transmitted from COSMOS satellites—are rarely copied in the western hemisphere because the satellites are often turned off by Soviet ground controllers prior to loss of signal (LOS). European stations receive the pictures regularly when a COSMOS satellite is operational. US stations hear them on occasion when the satellites are left on to support Soviet fleet maneuvers in the North Atlantic and North Pacific areas, and on other occasions where they remain on either accidentally, or as a result of changes in Soviet ground-station operations.

GEOSTATIONARY SATELLITES

Geostationary satellites are in circular orbits over the equator at an altitude of approximately 22,000 miles. At this altitude, a single orbit of the earth takes 24 hours—precisely the time required for the earth to rotate once beneath the satellite. Thus, while the satellites are moving, the earth rotates below them at the same angular rate. In effect, the satellites remain over the same point on the equator and, from the ground, they appear to remain at the same point in the sky. Once the proper antenna bearing for a specific satellite has been determined, the antenna can (except for some occasional adjustments) simply be locked in place for the operational lifetime of the satellite in question. The satellite signals are beamed back to earth at microwave frequencies, so a small parabolic dish antenna is usually used in conjunction with a suitable converter ahead of the station receiver.

The majority of the operational geostationary weather satellites have the primary mission of imaging their hemisphere using a multispectral scanning radiometer that provides very-high-resolution images at both IR and visible wavelengths every 30 minutes. The satellites spin on their axes at approximately 100 revolutions per minute (r/min), providing the horizontal scanning, while a motorized mirror with a period of approximately 20 minutes is used to provide the vertical scanning. This data is relayed back to earth in a very-high-density digital format that requires specialized equipment for reception and display. This image data is processed by high-speed computers on the ground that provide two different functions. First, the original data, obtained during the 30 ms of each 600-ms horizontal scan, is retransmitted in "stretched" form (at a lower bandwidth) during the 570 ms when

USA NOAA GOES-E 02/22/87 1800Z NE VS

Figure 1.7—An example of a GOES-E NE visible-light quadrant. This area of the earth was imaged at the same date and time as the IR example in Figure 1.6, but this sample shows the visible-light data. Visible-light quadrants such as this one can be recognized immediately by the fact that the view of space beyond the limb of the earth is black. Note, however, that there are many clouds in this view that were not obvious in the IR version in Figure 1.6. There is a major extension of the cloud system over the southeastern US extending across the Gulf to Yucatan. Only a few bits of this extension are evident in Figure 1.6. These clouds must therefore be quite low and warm, thus lacking contrast against the warmer Gulf waters. Similarly, clouds blanket Nicaragua and Panama, but are only slightly evident in the IR view for a similar reason. The same is also true for small cloud elements associated with Cuba and other islands in the Caribbean. A detailed comparison of visible and IR data can thus yield information on the vertical distribution of weather systems.

the satellite sensors are scanning empty space. This signal is in a digital format, and although a number of amateur stations have succeeded in displaying this data, such a project is still too exacting and/or expensive for widespread use.

The same ground computers also sector the data into individual quadrants, then relay these images in analog form back through the satellite as part of the Weather Facsimile (WEFAX) program. WEFAX transmissions use the same AM subcarrier as the polar-orbiter transmissions; the WEFAX signal format is described in detail in Chapter 4.

US WEFAX operations all originate from the GOES (*Geostationary Operational Environmental Satellite*) satellite series. The GOES network typically consists of three satellites. GOES-E, stationed at 75° W longitude, and GOES-W at 135° W longitude, are imaging satellites that transmit WEFAX pictures between primary image acquisitions. The third satellite, GOES-C, is situated at 107° W longitude and functions entirely as a WEFAX relay. GOES-C is usually an older satellite, retired from either the GOES-E or GOES-W position, and carries both GOES-E and GOES-W products. The positions given here are the nominal ones for a com-

plete three-satellite network. Unexpected failure of one or more of the primary satellites can cause operational positions to be changed—sort of a space ballet—as functional satellites are nudged along the geostationary track to take up new positions to help compensate for the loss of a satellite.

The US primary GOES data is sectored in several ways for WEFAX transmission. The primary mode involves breaking the disc into four quadrants: northwest (NW), northeast (NE), southeast (SE) and southwest (SW). In addition, a tropical east (TE) and tropical west (TW), centered on the equator, are also available.

WEFAX transmissions derived from GOES data are of two types, representing visible-light and IR data. During the day, transmissions of visible-light quadrants are available. Grids outlining geographic and political boundaries, and longitude and latitude references (all added by the NOAA ground computers), make it easy to locate specific features (Figure 1.6).

Visible-light data formerly was not gridded in this fashion (Figure 1.7). With proper attention to the daily transmitting schedule, matching sets of visible or IR quadrants can be obtained, making it possible to

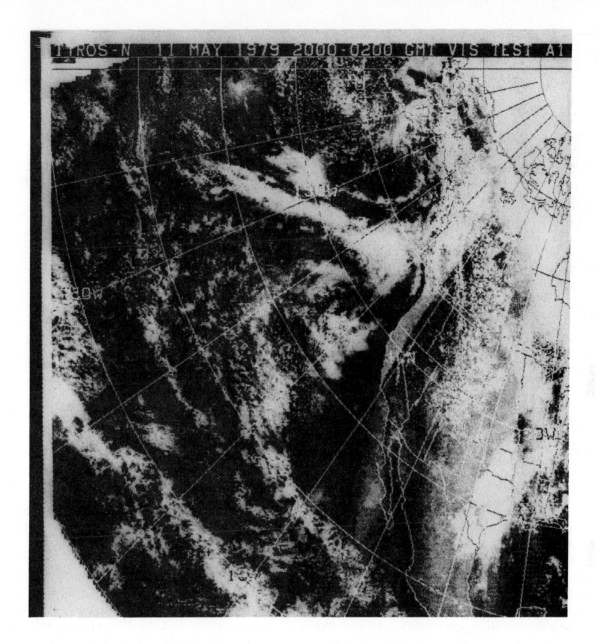

Figure 1.8—An example of a polar mosaic, prepared from TIROS visible-light data by the satellite-service ground computers and transmitted via GOES WEFAX. In this view, the north pole is at the upper right and out of the picture. This view covers western North America and a considerable area in the eastern Pacific, and was compiled from TIROS visible-light data. The image was printed on a version of the fax recorder illustrated in Chapter 4, using electrostatic

reconstruct the entire earth disc.

If you observe visible-light and IR images from the same quadrant, obtained at the same time, they rarely look identical. First, the view of space beyond the limb of the earth appears black in the visible-light image, but white (cold) in the IR image. Quite a few cloud features that are apparent in the visible-light image may appear to be lacking in the comparable IR view. This is because IR-image contrast is a function of

temperature differences. Cloud features close to the ocean surface are clearly evident in visible light, but because these low clouds are relatively close to land or sea surface temperature, they may be obscure or absent in the IR data. High, cold clouds appear very white in IR as opposed to lower (and warmer) clouds that appear as mid-range grays in IR images. These variations in tonal values allow you to reconstruct the vertical distribution of cloud systems in IR pictures,

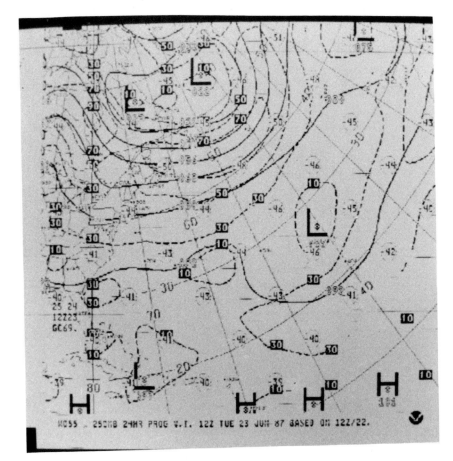

Figure 1.9—An example of a weather chart transmitted via GOES WEFAX. In this case, the chart is a polar projection (pole off to the upper left) covering the eastern US and Canada and most of the north Atlantic. This particular chart is a 24-hour VT (vertical temperature) prognosis. Chart ID data are contained in a footer along the bottom of the image.

while a comparison of comparable visible and IR data provides the best overall view of weather systems.

In addition to the primary GOES image data, WEFAX is also used to transmit computer-generated Mercator or polar mosaics generated from visible and IR TIROS/NOAA polar-orbiter data (Figure 1.8). Because these products cover the entire world, it is possible to follow weather developments almost anywhere, regardless of your station location. WEFAX is also used to transmit weather charts (see Figure 1.9) as well as the WEFAX transmission schedule and polar-orbiter TBUS messages (see Chapter 8).

The European Space Research Organization operates a geostationary satellite, METEOSAT, (positioned at 0° longitude), which is quite similar to the US GOES operation in a number of respects. In the case of METEOSAT, the earth disc is broken down into a larger number of smaller quadrants. Although GOES WEFAX operations are confined to a single frequency (1691 MHz), METEOSAT operates on 1691 and 1694.5 MHz. The Japanese operate a similar satellite (GMS) over the eastern Pacific, but the wider RF bandwidth of these transmissions requires the use of larger antennas (see Chapter 3). Although the GOES WEFAX service was initially viewed as too complex for amateur use, advancements in technology and exten-

sive amateur experimentation have made WEFAX accessible to almost any installation at relatively modest cost.

BASIC GROUND-STATION COMPONENTS

Most newcomers to weather-satellite activities start by assembling a polar-orbiter receiving installation. These satellites transmit on frequencies in the 136- to 138-MHz range, and require nothing more than a small VHF antenna (Chapter 2) and a simple VHF FM receiver of suitable design (Chapter 3). The picture-display system can take many forms, and several alternatives are discussed in Chapter 4, with a complete digital scan converter described in Chapters 5 and 6. Options for computer interfacing, to enhance the capabilities of the system, are described in Chapter 7. In addition, you'll have to learn techniques for predicting satellite orbits to know when to expect passes from a given satellite and these are covered in Chapter 8. Integrating all of this into the overall operation of your receiving station is the subject of Chapter 9. Advanced techniques, such as image processing, are discussed in Chapter 10.

Upgrading to WEFAX requires the addition of a small microwave antenna (Chapter 2) and a down-converter (Chapter 3) to convert the microwave GOES signals to the 136- to 138-MHz VHF range. In most cases, a display system suitable for use with polar-orbiter pictures can be used directly or adapted for WEFAX display. I'll describe this display system in Chapters 5 and 6. It is completely compatible with WEFAX and all the present polar-orbiter modes. Antenna positioning calculations are covered in Chapter 8.

How much you spend on your satellite installation depends on which satellite signals you want to receive, how many features you desire, how much of the station you are willing to build, and whether or not you are adept at combing flea markets and surplus outlets for bargains. Completely functional polar-orbiter installations have been built for science-fair projects by Junior High students for as little as $200. At the other end of the scale, you can easily invest several thousand dollars if you buy everything assembled and tested and insist on having the latest model of every possible piece of equipment.

Fortunately, this hobby is one in which some of your own skills and ingenuity can compensate greatly for a lack of ready cash. No matter what the cost of your station and no matter how simple or sophisticated it might be, there is endless potential for fun, fascination, and the daily excitement of watching the earth from space.

Chapter 2

Weather-Satellite Antenna Systems

INTRODUCTION

Your antenna system will consist of two primary elements. The first is the antenna, which is designed to intercept the small amounts of RF energy reaching your location from the distant satellite. The second system component is the transmission line, a cable designed to carry these faint signals from the antenna back to your receiver, hopefully while minimizing signal losses in the transfer. Most of this chapter is devoted to the antenna portion of the system, but I'll have a general discussion of transmission lines and the all-important cable connectors at the end of this chapter.

Many weather-fax enthusiasts will find the installation of an antennas (or antennas) to be the single greatest problem they have to solve in getting a weather-satellite station operational. Increasing numbers of condominiums, urban apartments, and suburban neighborhoods with restrictive real-estate covenants may make it appear that there is no option for an effective weather-satellite antenna system. If this is your situation, there is hope. A new section, **OPTIONS**, has been added at the end of this chapter. In it you should find one or more solutions that will work for you no matter how restrictive your living situation might be.

ANTENNA BASICS

Gain

Gain is a measure of how much a given antenna increases the level of a signal relative to some reference standard—usually a simple dipole. The gain units are decibels (dB) and, since the scale is logarithmic, a gain of 3 dB equates to a doubling of the signal strength. Gain is highly desirable since it helps to overcome cable losses and makes the RF preamplifier in the receiver less critical. Unfortunately, to get an increase in gain you have to trade off another antenna parameter—the beamwidth.

Beamwidth

Beamwidth is a measure of the width of the antenna pattern. Generally, an antenna with low gain has quite

a wide pattern, receiving signals well from a number of different directions. If we want to increase gain, we must do it at the expense of decreasing the beamwidth. A high-gain antenna must be pointed rather accurately, otherwise the received signal will be quite a bit weaker than would have been the case with a low-gain/wider-beamwidth antenna. A high-gain VHF antenna for one of the polar-orbiter satellites will easily deliver a very strong signal to the receiver, but to do so, it must be accurately tracked during a pass so that it remains pointing at the satellite at all times. In contrast, an omnidirectional antenna designed to accept signals from all directions may deliver a comparatively weak signal, but does not require tracking the satellite!

Polarization

The orientation of radio waves in space is a function of the orientation of the elements of the transmitting antenna. In our daily lives, we encounter two principal modes of polarization—*horizontal* and *vertical.* FM broadcast and TV transmissions are typically horizontally polarized, and an antenna designed for such signals has elements oriented horizontally. In contrast, police, other public service, and mobile (cellular) phone transmissions use vertical polarization because a simple vertical whip antenna is the easiest sort of omnidirectional antenna to mount on a vehicle. If a horizontally polarized signal is received on a vertical antenna, or vice versa, we refer to this as *cross polarization.* Cross polarization is highly undesirable since signal losses can exceed 20 dB—more than enough to render an otherwise strong signal completely unreadable.

Both vertical and horizontal polarization are examples of *linear* polarization and present real problems with space communications—particularly in the case of polar-orbiting satellites. Such satellites are in constant motion with respect to the ground station. Consequently, a linearly polarized signal from a satellite appears to be constantly changing polarization with respect to the ground station. The result can be strong signals at one time and very weak or unreadable signals just minutes later. Because linearly polarized antennas are simple and easily mounted on spacecraft, early satellites used them; the Meteor satellites still do. It was therefore necessary to solve the variable polarization problem at the ground-station end: The answer is to use *circularly* polarized antennas.

A circularly polarized wave rotates as it propagates through space, and antennas can be designed for either right-hand circular (RHC) or left-hand circular (LHC) polarization. An RHC- or LHC-polarized antenna shows a maximum loss of only 3 dB when receiving a linearly polarized signal, so circularly polarized antennas are almost universally used at polar-orbiting-satellite ground stations. Either an RHC- or LHC-polarized antenna would do fine for Meteor reception, but if we are going to also receive the TIROS/NOAA satellites, we must use an RHC-polarized antenna because the TIROS satellites generate an RHC-polarized signal—not a linearly polarized one. If we receive the TIROS signal on an RHC-polarized antenna, we experience no polarization loss. Should we try to receive the RHC-polarized signal using an LHC-polarized antenna, we would experience cross-polarization and have a constant 20-dB polarization loss to contend with. Thus, our best bet is to use an RHC-polarized antenna where we'll have no TIROS polarization losses and a maximum of 3 dB of Meteor signal loss.

A linearly polarized antenna can be used with the RHC-polarized transmissions from NOAA satellites with a maximum polarization loss of only 3 dB. Although such an antenna is much simpler than an equivalent one designed for circular polarization, it is unsuited for use with Meteor satellites. Some additional options with regard to antenna polarization are discussed at the end of the *Gain Antenna* section.

Geostationary satellites such as GOES do not move in relation to the ground, so linear antenna polarization can be used at both ends of the circuit.

Unfortunately, the two major frequency ranges we must use—VHF for polar orbiters and S-band for geostationary satellites—require that we use quite different antennas for each service. The VHF antennas are much like conventional TV antennas in terms of their design, but S-band antennas are almost always of the parabolic dish type, similar to those used for satellite-TV reception. There are other types of S-band antennas, but dishes are simpler to work with unless you have access to some pretty sophisticated test equipment.

VHF ANTENNA SYSTEMS

One of the most common forms of VHF antenna is the *Yagi*, named in honor of the Japanese scientist who first elucidated the principles of combining a basic dipole—a so-called *driven element*—with a number of *parasitic elements*. Although the driven element is connected to the transmission line, the parasitic elements are coupled to the driven element by resonance, not by any physical connection. Parasitic elements are of two general types. *Directors*, which are electrically shorter than the driven element, are placed in front of the driven element, facing the source of RF energy in the case of a receiving antenna. Most Yagis also employ a single *reflector* that is electrically longer than the driven element and is placed behind it. The driven element, in combination with one or more parasitic elements, makes up a beam antenna, of which your common TV antenna is a prime example. A simple beam usually starts with a driven element and a single reflector. As parasitic elements are added (usually in the form of an increasing number of directors) gain increases, beamwidth decreases, and generally, the frequency range over which the antenna will operate (bandwidth) becomes narrower, although this is far less serious for receiving applications than it is for transmitting.

To achieve circular polarization, a beam antenna can be built using two identical sets of elements, each mounted at right angles to the other. The two driven elements must be properly phased with a length of transmission line between them and, depending upon how the connections are made, either RHC or LHC polarization can be achieved. Commercial beam antennas are available from a number of manufacturers who are listed in Appendix I. You can also construct your own beam antenna using design formulas and construction techniques included in *The ARRL Handbook* (American Radio Relay League, see Appendix I). The four-element Yagi described in the *Gain Antenna* section is one easy approach to home construction. There are other types of directional VHF antennas that are documented in a variety of ARRL publications.

Any kind of directional antenna requires tracking to keep the antenna pointed at the satellite as it moves across the sky. This requires two antenna rotators—one to control the direction (or azimuth) in which the antenna is pointed and the second to control the elevation or the angle at which the antenna is pointed relative to the horizon. This hardware requires that you use a manual plotting board or a computer to determine the precise times for various azimuth and elevation settings to keep the antenna pointed at the satellite during a pass.

For most of us, the use of a directional antenna means we are confined to weekend and evening passes when we can operate the tracking system. There are approaches that can be used to automate the tracking process to permit unattended recording of satellite passes, but they'll increase the overall cost of your station. The popularity of Amateur Radio satellite communications has resulted in a number of manufacturers producing sophisticated rotators and control systems for moving antennas in both azimuth and elevation. Some of the systems can even be controlled by a remote computer. Such a rotator control system costs about $500, plus the cost of the computer system—if you want to take advantage of all of the flexibility such a system can provide.

For many operators, the antenna needs for VHF operation can be met with an omnidirectional antenna system. These antennas are simple and inexpensive to construct and have no need for azimuth or elevation rotators. Such an antenna will deliver noise-free pictures for the best passes of a given satellite (which occur twice each day), including the "worst-case" situation where a set of passes straddles the ground-station location. The major advantage of such an antenna system is that no tracking is required and unattended operation can be as simple as connecting a timer to the station tape recorder.

Unless you have the need to regularly access passes at extreme range, there is really no need for a beam antenna with all the additional complexities that are involved. Most of the TIROS/NOAA and Meteor images in this book were obtained using the omnidirectional antenna to be described in the next section, and graphically illustrate its effectiveness. In the case of overhead passes, I can hold the signal at full quieting from Ellesmere Island, north of Hudson Bay, to just north of Yucatan. In the case of worst-case straddling passes, the coverage is reduced from central Hudson Bay to just south of Florida.

AN OMNIDIRECTIONAL VHF ANTENNA

When I published an article, shortly after the appearance of the first edition of the *Handbook*, which I titled "An Omnidirectional Circularly-Polarized Antenna for Weather-Satellite Reception," a practical editor at *73 Magazine* shortened the title to the "Satellite Zapper." Despite my best efforts to return to the pompous original, it remains to this day as the "Zapper." The Zapper is simple in concept in that it is nothing more than a short beam (two driven elements and two reflectors) with such a wide beamwidth that, when pointed straight up, it functions as an omnidirectional antenna system that does not require any tracking for passes within your "best-pass" window. With a hot receiver and relatively short transmission line, the Zapper performs quite well, although it is at its best when a low-noise preamp—either a JFET (*junction field-effect transistor*)—or, better still, a GaAsFET (*gallium arsenide field-effect transistor*) is mounted at the antenna. With a preamp in place, the length of transmission line between the antenna and receiver is basically irrelevant.

The Zapper is an ideal first project because of its simplicity of construction and ease of mounting. If you later use a gain antenna to maximize your coverage at the limits of reception, you'll still use the Zapper regularly for general monitoring, spotting new Soviet satellites by scanning various frequencies, etc.

A few words might be in order for those of you who are familiar with VHF-antenna design. If you are a purist, you should be prepared to be horrified by some aspects of the Zapper's design. You may be tempted to "clean up" the design by following accepted rules for such things as matching, and so on. Please keep in mind that we are not dealing with transmitting antennas; instead, we're looking for the best possible reception with the simplest possible approach to getting the job done. The earliest pre-Zapper started out following all the rules—and it didn't work particularly well. With each revision, the design became simpler and performance increased. The newest version shown here is the simplest and best yet. You can clean it up if you want to, but be advised that it may not work as well as this one. The evolution from orthodox to unorthodox has been quite purposeful, and you should keep that in mind before embarking on major "improvements."

The Zapper requires a vertical mast, two reflectors, and two driven elements. The driven elements mount at right angles to each other at the top of the mast, offset vertically by a distance of two inches. The reflectors mount approximately 1/4 wavelength below the driven elements, with each reflector parallel to a driven element. The overall antenna is quite compact and unobtrusive. This is an important factor in many areas where restrictive deeds and real-estate covenants can severely restrict the kind of antennas that can be erected.

Materials

The original Zapper, which appeared in the second and third editions of the *Handbook*, was fabricated from various sizes of aluminum tubing. Such tubing can be hard to obtain and is relatively expensive from most sources. Assembling the antenna required careful drilling of holes that never seemed to line up properly, and the hardware always seemed to corrode, no matter how carefully the antenna was weatherproofed. For this edition, I redesigned the Zapper (let's call it Zapper II) so that it requires no aluminum, no drilling, no assembly hardware, and it won't corrode. It works as well as the original, costs less, and looks quite a bit better hanging up in the breeze.

The secret to this new version of the antenna is the use of standard 1/2-inch CPVC plumbing pipe and fittings that you can obtain from almost any hardware or discount store. (From here on, I'll refer to CPVC simply as PVC.) The antenna-element housings and their supporting framework are constructed entirely of PVC pipe and fittings, providing both the rigidity required and complete weatherproofing of the actual

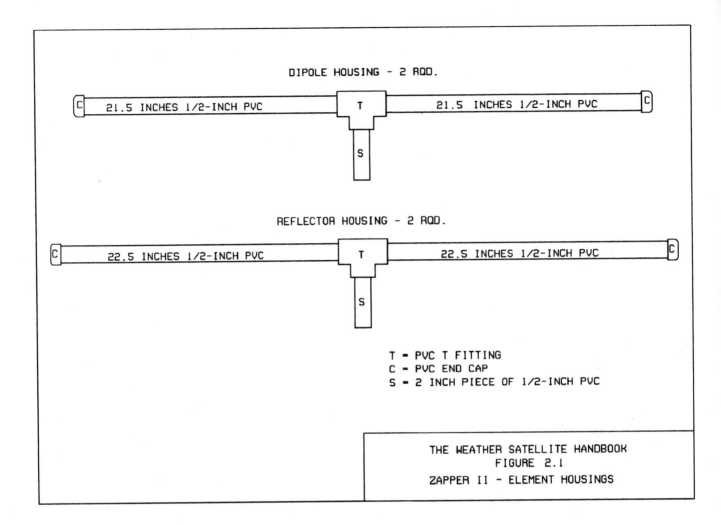

DIPOLE HOUSING - 2 ROD.

| C | 21.5 INCHES 1/2-INCH PVC | T | 21.5 INCHES 1/2-INCH PVC | C |

S

REFLECTOR HOUSING - 2 ROD.

| C | 22.5 INCHES 1/2-INCH PVC | T | 22.5 INCHES 1/2-INCH PVC | C |

S

T - PVC T FITTING
C - PVC END CAP
S - 2 INCH PIECE OF 1/2-INCH PVC

THE WEATHER SATELLITE HANDBOOK
FIGURE 2.1
ZAPPER II - ELEMENT HOUSINGS

antenna elements. To build the antenna you'll need the following materials:

2   10-ft (3 m) lengths of ½-inch (1.27-cm) PVC pipe
8   PVC **T** fittings for ½-inch pipe
11   ½-inch PVC end caps
1   bottle of PVC pipe cement compatible with your pipe
1   5-to 6-foot (ca. 2-m) length of aluminum TV mast
4   stainless-steel hose clamps

In addition, you'll require the following antenna items from your local electronics outlet or mail-order outlet:

1   20-ft length of RG-58 coaxial cable (Belden 8219 or equivalent), plus enough additional cable for the run from the antenna to your station location
1   8-ft length of 300-ohm TV twinlead
2   double-female BNC adapters

5   BNC plugs for RG-58 cable and an additional connector to match your receiver antenna-input jack
1   BNC **T** adapter (female arms, male common)

Use a hacksaw to cut the following lengths of ½-inch pipe:

3   1 inch (2.6 cm)
4   2 inch (5.2 cm)
4   21.5 inch (54.5 cm)
4   22.5 inch (57 cm)
1   8 inch (20.32)

Use a file to deburr the cut ends of the tubing and put them aside with the PVC fittings.

Reflectors

We'll start antenna assembly with the reflector elements because they're the simplest to fabricate. In the steps that follow, temporarily assemble the indicated pieces and check to make sure you have them properly aligned. When you're ready to make the assembly permanent, coat one end of the indicated piece of

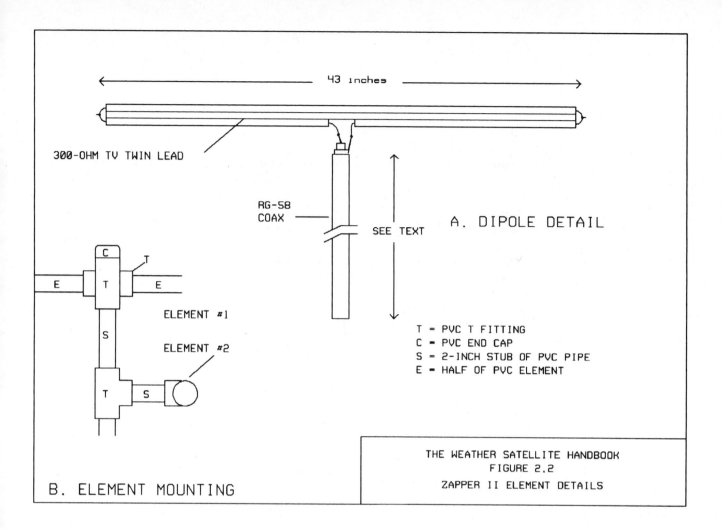

300-OHM TV TWIN LEAD

RG-58 COAX

SEE TEXT

A. DIPOLE DETAIL

43 inches

T = PVC T FITTING
C = PVC END CAP
S = 2-INCH STUB OF PVC PIPE
E = HALF OF PVC ELEMENT

ELEMENT #1

ELEMENT #2

B. ELEMENT MOUNTING

Figure 2.2—Zapper II element details. At A, one conductor of a piece of twinlead has been cut and soldered to a length of coaxial cable. At each end of the twinlead, the two conductors are twisted together and soldered (see text). At B, the method of mounting two elements at right angles to each other. In this view, you're looking at the back of the top **T** (element #1), with a PVC pipe cap at the top. A two-inch stub connects the lower end of the top **T** to a second **T**. The second **T** is attached to a third **T** by another two-inch stub. The third **T** (part of element #2) is viewed end-on from the capped end of the element.

PVC tubing with the tubing cement and insert the piece firmly into the indicated fitting.

Using the reflector housing diagram in Figure 2.1, cement two 22½-inch pipe sections into the side arms of one of your PVC **T** fittings. Cement a 2-inch length of pipe into the lower position of the fitting, then cement a PVC cap onto one end of the housing, leaving the other end open. Repeat this entire assembly sequence with a second set of pieces and set both reflector housings aside for at least one hour to let the cement cure.

Cut two 45-inch lengths of RG-58 coaxial cable. When the reflector housings have set, insert one piece of the cable down the length of each housing, cement a PVC cap to the open end, and lay the assembly aside.

The following assembly steps refer to Figure 2.2B. Insert a 1-inch length of tubing into one end of the cross-arm of a **T** fitting and cement a cap to the end of this stub. Insert a 2-inch piece of pipe into the opposite end of the fitting. Insert the pipe stub from one of the reflector housings into the side arm of this fitting so that the element is at right angles to the long axis of the fitting, as shown in the upper part of Figure 2.2B. Your orientation should be accomplished quickly because the cement sets rapidly.

Add a second **T** fitting to the free end of the 2-inch stub, oriented at right angle to the upper fitting. Insert a 1-inch length of pipe into the bottom of this lower fitting and cap it. Now insert the 2-inch stub from your second reflector housing into the side arm of the lower **T** fitting. You should end up with your two reflector housings at right angles to each other and separated by about 2 inches. Lay aside the complete reflector assembly, but keep it near at hand so you can refer to it as you assemble the driven-element housings.

Driven Elements

Construction of the two driven elements begins with the assembly of two dipoles, illustrated in Figure 2.2A. Cut two 4-foot (1.2-m) lengths of RG-58 cable. Carefully cut through the plastic jacket at a point about 2/3 inch (2 cm) from one end of each piece. Score the jacket lengthwise from this point and peel the jacket away. Unravel the shield braid, then twist it together into a single, short stub. Remove the insulation from the upper half of the exposed center conductor. Use a soldering iron to tin both the center conductor and the tip of the braid stub.

Cut two 44-inch (111.8-cm) lengths of the 300-ohm TV twinlead. For each piece, cut enough of the webbing from between the conductors at each end so that you can strip the two conductors, twist them together, and solder the connection. Mark the middle of each piece and cut through one of the conductors at that point. Strip 1/2 to 1 inch (1 to 2 cm) of insulation from the wires on each side of the cut. For each piece, solder the center conductor of one of the pieces of coax to one wire and the braid to the other. You should now have two dipole assemblies like the one illustrated in Figure 2.2A.

Dipole Housing

We'll now build the PVC housing around each of the dipoles you have just constructed. Feed the two ends of the dipole through the base of a T fitting, routing one side of the dipole out one side arm opening and the other out through the remaining opening. Work the individual dipole legs outward from the fitting until the solder connections to the coaxial cable are located up inside the T fitting. Slide one 20-inch piece of pipe over one leg of the dipole and cement it to the T fitting, repeating the operation with the other dipole leg and a second 21½-inch piece of tubing. Cap the open ends of the tubing. Now take a 2-inch length of pipe, slide it up the coax and cement it to the remaining opening in the fitting. You should now have a pair of dipole assemblies, enclosed in the PVC pipe with the coaxial cable coming out the 2-inch pipe stub on each housing.

All that remains now is to use the remaining caps and T fittings to make a complete driven element assembly (two elements) exactly as you did with the reflectors. The orientation of all the pieces should be just the same as the reflector assembly. Because the lengths of coax have to be threaded through the T fittings, I suggest the following assembly sequence:

1) Thread the free end of the coax from one dipole through the side hole of a T fitting (assuming the final vertical orientation of the T fitting) and out the lower hole.

2) Cement the 2-inch dipole stub to the side hole, orienting the fitting at right angles to the dipole assembly.

3) Repeat the preceding two steps with the second dipole assembly and the remaining T fitting.

4) For the uppermost dipole, insert a 1-inch pipe stub in the upper hole and cap it.

5) Slide a 2-inch piece of tubing up the coax of the upper dipole and cement it to the lower hole of the T fitting where the coax exits.

6) Feed the upper coax cable through the upper hole of the lower T fitting and out the bottom hole with the coax from the lower dipole, cementing the pipe from the upper fitting to the upper hole of the lower fitting. Be sure the two dipoles are at right angles to each other before the cement has a chance to set.

7) The lower T fitting should now have two cables exiting from the lower hole. Slide the 8-inch pipe piece up both cables and cement it to the lower T fitting.

After the dipole assembly has set, install BNC connectors on the free ends of the two lengths of coax. One cable (the one from the lower dipole) will appear longer than the other. Do NOT trim them to equal length when installing the connectors!

Cut a 15-inch length of coaxial cable and install BNC connectors on each end. This piece of cable is our phasing line.

Final Mounting

Stainless-steel hose clamps are used to mount the dipole and reflector assemblies on the length of aluminum pipe or TV mast. If circumstances permit, it's usually easier to mount the assemblies with the mast in place. Alternatively, you can mount them on the mast and then install the mast section. The dipole assembly should go at the top of the mast, with the ends of the upper dipole facing E-W. The reflector assembly should be mounted so that the upper reflector is parallel with the upper dipole and 17 inches (44 cm) below it. Your mounted antenna should look similar to that shown in Figure 2.3—but with the reflector and driven elements properly oriented!

For the best possible results, the antenna should be mounted as high as possible with a minimum of obstructions within 5 degrees of the horizon. If a preamplifier is mounted at the antenna, the length of transmission line between the antenna and receiver is irrelevant, allowing you to choose the best possible site for the antenna without worrying about transmission-line losses.

Figure 2.3—A photograph of the prototype of the Zapper II antenna. The antenna is quite unobtrusive and can be installed using a small tripod or other minimal support. You should take care where the assembly notes indicate that the reflector and driven element assemblies have the same orientation. With this prototype I didn't, and ended up with the lower reflector on the other side of the mast from the lower driven element. Although this should skew my pattern a bit to the east in one plane, there doesn't seem to be any obvious effect.

Cables

The longer cable from the dipole assembly should be connected to one arm of the BNC **T** connector. Connect the phasing line to the other arm and use a double-female BNC connector to connect the free end of the phasing line with the shorter cable from the dipole assembly.

For the best possible performance, the preamplifier should be mounted at the antenna with a housing to keep rain or snow off the unit. A suitable housing can be made from a plastic freezer container, secured to the mast with hose clamps. The top of the container should face down and the cover should be equipped with holes to let the cables pass up from below. These holes also vent the case so that condensation does not build up inside the enclosure. The male connector of the BNC **T** fitting can connect directly to the pream- plifier input, or you can use a short length of intercon- necting cable. The transmission line to the station receiver can connect directly to the preamplifier out- put. If your preamplifier is powered through the trans- mission line, your installation is now complete. Otherwise, you'll need to run a wire with the transmis- sion line to carry 12 V dc to the preamplifier. A ground wire is not required because the coaxial cable shield provides the ground return.

Although the antenna performs best with a mast- mounted preamplifier, a direct run of cable (using another double-female BNC adapter) can be con- nected to the BNC **T** connector. For best results, the receiver should have a low-noise front end, and no more than 50 feet (15 meters) of transmission line should be used—preferably less.

Finishing Touches

Prior to taking care of the final details, run a recep- tion check on the antenna system. Using a vintage Vanguard receiver and a Hamtronics GaAsFET pream- plifier, I can expect to hear a satellite on the horizon with full quieting by the time the satellite has reached an elevation of 3 to 4 degrees, assuring a minimum of 12 to 14 minutes of coverage for a worst-case situation with a pair of straddling passes. With the exception of a few examples of West Coast passes, all the polar- orbiter images in this book were obtained with the omnidirectional antenna system.

As a final check on polarization, observe a number of NOAA passes. If there are deep nulls in the pattern, as evidenced by noise at moderate to high satellite elevation, switch the phasing line position from the short to the long cable.

Once you know the antenna is working, use plastic electrical tape to cover all exposed connectors. Once they have been taped, cover them with the putty-like sealer available at most electronics shops that carry TV or CB antenna supplies, and tape them again. The sealer can also be used to seal the base of the dipole assembly where the cables exit. Electrical tape can then be used to secure the cables neatly to the mast. A coat or two of paint will protect the tubing from the effects of UV radiation and assure an almost unlimited service life for your antenna assembly.

One area where you can experiment concerns the spacing between the driven element and reflector as- semblies. The 17 inches (57 cm) specified represents approximately 0.2 wavelength at 137 MHz, and is a good starting point. Increasing this spacing changes the pattern slightly in ways that might prove useful to you. At a maximum spacing of ⅜ wavelength (32.25 inches [82 cm]), the gain directly overhead is de- creased slightly, but gain improves closer to the hori-

Figure 2.4—The WEFAXTENNA from Vanguard Labs, an example of a moderately priced commercial omnidirectional antenna. This antenna uses a pair of phased dipoles over a ground plane consisting of 8 radial elements. The mast is a piece of thick-walled PVC tubing that provides weather protection for the dipole phasing lines as well as the preamplifier which is powered through the transmission line. The antenna is available with or without the preamplifier.

zon. You should not exceed this value of ⅜ wavelength because beyond that point, the pattern breaks down from a single broad lobe to a series of minor lobes that will lead to very erratic reception.

Commercial Options

Omnidirectional antennas similar to the Zapper have now become quite fashionable and a number of vendors offer designs that represent a viable alternative to building your own. Most feature integrated weatherproof preamps as options or as standard equipment. Figure 2.4 illustrates the WEFAXTENNA, an omnidirectional system marketed by Vanguard Labs. This is a moderately priced system requiring only an hour or so of assembly. It uses an integrated RF preamplifier and produces results comparable to the Zapper home-built antenna.

Although the Zapper gives a good account of itself for the best passes of the day, tracking a satellite out to the eastern or western horizon requires a gain antenna system. Figure 2.5 (which also appears on the cover of this edition) is a mosaic prepared from two passes from NOAA-11 on 2 Sep 1989. The eastern pass, extending from northern Yucatan to northern Hudson Bay, was a near-overhead pass slightly to the east and was logged with the Zapper antenna. The western pass is another story, for the maximum satellite elevation was about 13 degrees. Although the Zapper would have provided perhaps 6 minutes of coverage for that pass, the use of a small 5-element beam permitted me to get over 12 minutes of coverage, resulting in a spectacular mosaic of most of North America.

When receiving satellite images first became popular, the VHF antenna arrays tended to be fairly complex, involving at least 10 vertical and 10 horizontal elements phased for circular polarization. The problem in those days was to get enough gain to overcome the poor noise figures of the available RF preamplifiers. Such antennas had very sharp patterns that required accurate tracking of the satellite to take advantage of the gain provided by the antenna system. Today, with the universal use of GaAsFET preamplifiers, a far more modest antenna system yields horizon-to-horizon coverage with a far broader pattern, leading to a considerable relaxation of the required tracking accuracy.

A Basic Four-Element Beam

A four-element beam provides a useful basic building block for a variety of final antenna options. If you want to build such an antenna, it is possible to use the PVC pipe approach employed for the Zapper, or you can work up the same basic antenna using aluminum tubing and more conventional antenna construction. Let's look at the PVC version first.

The essential dimensions of the beam are shown in Figure 2.6A. The beam is constructed using PVC pipe and fittings, just like the Zapper. You'll need two 10-foot (3-m) pieces of ½-inch pipe, 8 **T** fittings, and 9 end caps. Cut your tubing into the following lengths: four 20½ inch (52 cm), two 21½ inch (54.6 cm), two 22½ inch (57.2 cm), two 16 inch (41 cm), one 22 inch (56 cm), and five 1 inch (2.5 cm).

Follow the Zapper instructions for the construction of one reflector and one driven element. The only deviation involves the use of 1-inch pipe stubs for these elements instead of the 2-inch units specified for the Zapper. This permits the **T** fittings of the elements to be mounted flush with the **T** fittings of the boom when we get to final assembly.

Figure 2.5—A two-pass NOAA-11 mosaic covering almost all of North America. The eastern pass was essentially overhead at my location and was logged with the Zapper II antenna. The western pass had a maximum elevation of 13° and was obtained by manually tracking the satellite with a small, 5-element Yagi antenna. Combining two passes like this is a bit trickier than simply linking the northern and southern pieces of a single pass. Note, for example, that the passes do not simply lie neatly side by side. In fact, they converge toward the poles, as your tracking exercises in Chapter 8 will indicate. Although you can expect to achieve a near-perfect match with geographic features, the match is likely to be a problem with clouds, especially toward the south. This is because the two passes are separated by about 102 minutes—almost two hours—during which the cloud systems will move! Differential lighting can also be a problem. Note that the region north of Lake Superior is quite dark in the eastern pass because of low light levels. The same latitude in the later western pass has more favorable lighting conditions. Such variations could be compensated for, to some extent, in the darkroom, but I did not make that effort in this case.

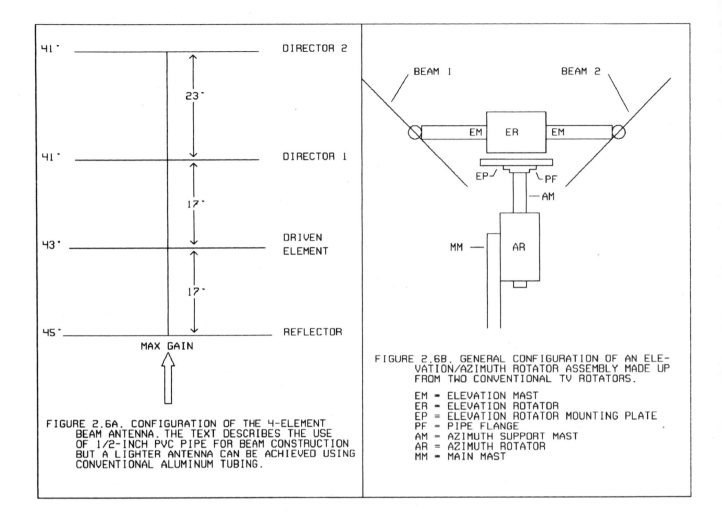

FIGURE 2.6A. CONFIGURATION OF THE 4-ELEMENT BEAM ANTENNA. THE TEXT DESCRIBES THE USE OF 1/2-INCH PVC PIPE FOR BEAM CONSTRUCTION BUT A LIGHTER ANTENNA CAN BE ACHIEVED USING CONVENTIONAL ALUMINUM TUBING.

FIGURE 2.6B. GENERAL CONFIGURATION OF AN ELEVATION/AZIMUTH ROTATOR ASSEMBLY MADE UP FROM TWO CONVENTIONAL TV ROTATORS.

EM = ELEVATION MAST
ER = ELEVATION ROTATOR
EP = ELEVATION ROTATOR MOUNTING PLATE
PF = PIPE FLANGE
AM = AZIMUTH SUPPORT MAST
AR = AZIMUTH ROTATOR
MM = MAIN MAST

You'll also need to construct two directors. These are assembled just like the reflector except that each uses a 41-inch (104-cm) length of coax, two 20½-inch (52-cm) pipe pieces, a **T** fitting, a 1-inch pipe stub, and two end caps.

The boom consists of four **T** fittings interconnected by PVC pipe sections. The reflector and driven element **T** fittings are interconnected by a 17-inch (41-cm) pipe section, the driven element and first director fittings are linked with another 17-inch (41-cm) pipe, and a 22-inch (56-cm) section links the fittings for the first and second directors. Insert a 1-inch (2.5-cm) stub at the front (second director end) of the boom and cap it. Mount the two directors and the reflector at right angles to the boom, then mount the driven element, threading the coax out the hole at the rear of the boom. Adding a connector to the cable and sealing the exit hole for the cable completes the beam—except for mounting hardware.

There are two basic disadvantages to the PVC version of the antenna. First, the boom has to be stiffened (see *Installation Options*) with a piece of aluminum angle stock. Second, the beam is relatively heavy compared to an aluminum version of the same antenna. These two disadvantages are offset by the ready availability of the materials compared to aluminum tubing.

The basic dimensions in Figure 2.6A can be used as a guide to duplicating the beam using aluminum tubing if desired. *The ARRL Antenna Book* shows several approaches to VHF Yagi construction and all the 2-meter antennas can be modified according to the dimensions provided.

Although several companies do market Yagi antennas cut for 137 MHz, most are somewhat expensive because they are produced in relatively small quantities. The same manufacturers typically market small Yagi designs for the amateur 2-meter (144 to 148-MHz) band. These antennas usually cost less because they are produced in much larger numbers and the manufacturers are competing for sales. What is generally not realized is that these 2-meter antennas will work almost as well (receiving, *not* transmitting!) as an antenna cut for the 136- to 138-MHz satellite band. A small 2-meter antenna will show good gain and directivity when used for weather-satellite reception, and the antennas are readily available from Amateur Radio supply houses throughout the world, as well as directly from the manufacturers. Typical models that do a good job in this service include:

Vendor	Model	Configuration	List Price
Cushcraft	A147-4	4-element linear	$50
	A144-10T	5+5 circular	$85
Hy-Gain	205S-1 25BS	5-element linear	$45
KLM	2M-4X/144-148	4-element linear	$62

If you are really interested in saving money, check with local hams. It is quite common for radio amateurs to have one or more old 2-meter beams stashed in the garage or other storage area. One of these old antennas, cleaned off with steel wool and with new element-mounting hardware, does an excellent job receiving weather-satellite signals—and you may be able to get the antenna for free!

If you ask around, you'll probably hear that 2-meter beams cannot possibly work in the 136- to 138-MHz range. The fact is that they do! The western half of the mosaic in Figure 2.5 was received using a *very* old 5-element 2-meter beam. As a new radio amateur (WV2EMC/WA2EMC) back in 1958, in northern New Jersey (living, believe it or not, in a log cabin on one of those little lakes that you can see in Figure 1.1), I was given the beam by W2KHQ so I could log onto the 2-meter Civil Defense net. The antenna is certainly over 35 years old, yet was still residing in the back of the garage after all those years and a great many intervening moves. Disassembled, cleaned, and with replacement hardware, the antenna works just fine!

Installation Options

The 4-element beam, as described, is a linearly polarized antenna with a gain of approximately 9 dB over a simple dipole. When used in any orientation to receive TIROS/NOAA signals, the actual gain is about 6 dB due to the 3 dB loss experienced when receiving the RHC-polarized satellite signal. This gain is entirely adequate. The various linear 2-meter antennas are roughly comparable. You can install such an antenna on a commercial azimuth-elevation (az-el) rotator system (such as the Yaesu G-5400B), or you can construct your own az-el rotator system using two standard TV antenna rotators.

The single antenna won't suffice for Meteor reception because these satellites use linear antennas. The easiest approach to achieving circular polarization is to build a second 4-element Yagi and orient it so that its elements are at right angle to the first Yagi. The phasing line and BNC **T** connector can then be used to interconnect the two antennas following the description for cabling the Zapper.

Figure 2.6B shows the general configuration for an az-el rotator system using two TV rotators. The azimuth rotator is mounted to your main support mast in the usual TV fashion, and its control box is used to control the antenna azimuth or compass direction, just as it would for normal TV service. The critical difference is that instead of turning the antenna directly using a short section of mast in the azimuth rotator, the rotator turns a short pipe stub with a pipe flange secured to a flat, metal plate. The second rotator, which is used for elevation control, is mounted, on its side, to this plate. A short section of mast from this second rotator mounts to the antenna, allowing that rotator control box to move the antenna up and down.

Many of the details of constructing your rotator system will depend on the TV rotators available to you. Most rotator housings are weatherproofed, but since they are not designed to be mounted on their sides, the sealing may not be completely effective for the elevation rotator. It may be worthwhile to fabricate a splash guard of sheet metal or plastic to keep the worst of the rain off the elevation-rotator housing. A rotator in which the movable mast extends completely through the housing is the best overall option because it permits the use of a counterweight to balance a single antenna, or lets you mount two antennas for circular polarization. A rotator without this feature can still be used for elevation control of a single antenna if the horizontal mast extension to the antenna is not more than about 2 feet (70 cm) long.

Mounting the PVC beam antenna requires that you stiffen the boom. This can be accomplished with a piece of 1/2-inch aluminum angle stock (cut to the length of the boom), using stainless-steel hose clamps to secure the boom along the length of the angle stock.

Mounting the antenna is done using an aluminum plate about 4 inches (10 cm) square. The plate should be fairly stiff so a thickness of at least 1/8 inch (3 mm) is indicated. Use a pair of **U** bolts to secure the boom to the plate at the balance point for the antenna— somewhere between the driven element and first director. Another pair of **U** bolts is then used to secure the plate to the mounting mast so that the boom is at right angles to the mast. If a single antenna is used, the elements should be oriented vertically to minimize the interaction with the mast and rotator. If you are mounting two antennas for circular polarization (necessary only for Meteor reception), mount each at 45 degrees to the horizontal mast and 90 degrees to each other, as shown in Figure 2.6B. If you are using one of the 2-meter antennas, they should already have mounting hardware installed, making it an easy job to mount one or two of them.

If your situation does not permit you to mount such an antenna system, or you are only interested in casual operation and don't wish to expend the time and money for a permanent installation, check out some of the suggestions in the **OPTIONS** section at the end of the chapter.

S-BAND GOES/METEOSAT ANTENNAS

The importance of adequate gain in the S-band receiving system will be highlighted in Chapter 3. In terms of antenna gain at these frequencies, the major determinant is the diameter of the dish. The job of the dish is to reflect the incoming RF energy to a focal point where it can be picked up by a feed horn or other device to transfer the RF energy to a transmission line. The bigger the dish, the greater will be the intercepted RF energy and hence, the greater the gain! Usable GOES/METEOSAT dishes come in quite a range of sizes:

Diameter			
Feet	Meters	Source	Gain (dBi)
2	0.6	Saucer sled	18
4	1.2	Commercial/surplus	24
6	1.8	TVRO	27.5
10	3.0	TVRO	32

At the small end of the size range (2-3 feet), we can have a workable and inexpensive antenna, but there are two major drawbacks. First, the minimal gain puts great demands on the RF preamplifier—and there is no margin for any appreciable line losses between the antenna and the preamp input. Secondly, the small antennas have such a wide pattern that at certain elevation and azimuth angles you may experience interference from adjacent satellites. I have a 2-foot dish that I use for demonstrations and portable work, but if I train this small dish on GOES W, I also get interference from GOES C! Still another problem with very small antennas is that if a problem develops with a satellite that results in a drop in signal level, you may no longer have enough of a signal margin for noise-free reception.

A 4-foot (1.2-m) dish (Figure 2.7) is about optimum for combining small size with adequate gain and a narrower beamwidth. Larger dishes will provide a greater gain margin or will permit you to use a less-effective preamp or, at the 10-foot (3-meter) end of the range, you can even get by with no preamp. Six-to 10-foot dishes are universally available from dealers in TVRO satellite equipment, and prices have become very competitive. Such antennas are available with either manual or remotely actuated polar mounts that permit you to swing across the entire geostationary arc—a real plus if you want easy access to all the GOES satellites. Such mounts are typically designed to mount on a pipe set into the ground in concrete but accessories for roof mounting are also available. If you buy a new antenna, get a quote on the system without the

Figure 2.7—Unlike TVRO satellite dishes, a WEFAX antenna need not seem particularly impressive. The difference in visual impact between this 4-foot WEFAX antenna and a typical 10- to 12-foot TVRO dish is considerable. The dish can be further blended into the background with the judicious use of paint if your living situation demands a low profile.

TVRO feed system or LNA. These items are of no use in our application and you should not be paying for them. Be aggressive in looking for the best price and you can save a great deal of money. Purchase of a used dish and mounting hardware from someone who has become disenchanted with satellite TV is still another option. Commercial surplus dishes are also available in some areas.

The primary disadvantage of large dishes is their narrow beamwidth. They must be aimed accurately—and if a satellite begins to drift near the end of its operational service, it is possible to get periods of the day during which the signal is degraded because the satellite drifts out of the antenna pattern.

Feed-Horn Design

The simplest approach to picking up the RF energy reflected from the dish is a feed horn, which is simply a metallic cylinder, closed at the far end. The cylinder acts as a waveguide at these microwave frequencies, and the RF energy inside the horn is picked up by a small probe coming in from the side (see Figure 2.8A). All of the dimensions of the horn, including its length (A), diameter (B), the distance between the probe and the closed end of the horn (C) interact to some extent and are dependent, along with the length of the probe, on the free-space wavelength at F, the fre-

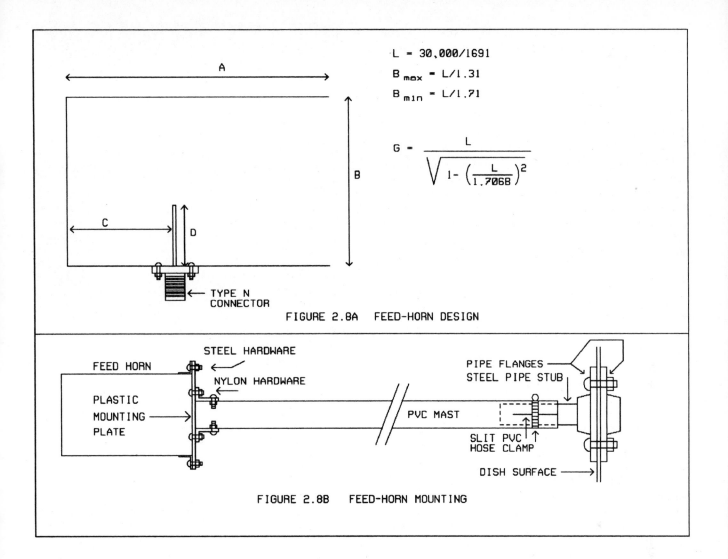

$$L = 30,000/1691$$
$$B_{max} = L/1.31$$
$$B_{min} = L/1.71$$

$$G = \cfrac{L}{\sqrt{1 - \left(\cfrac{L}{1.706B}\right)^2}}$$

FIGURE 2.8A   FEED-HORN DESIGN

FIGURE 2.8B   FEED-HORN MOUNTING

quency of interest (1691 MHz). All of the design equations are contained in Figure 2.8A, and I will not repeat them here. At 1691 MHz, the free-space wavelength is 17.74 cm, and that determines the minimum and maximum diameter of the horn (Bmin and Bmax) that will support propagation of the microwave energy within the horn. Our horn should have a diameter between 10.37 and 13.54 cm.

Once the microwave radiation is inside the horn, its wavelength changes to a so-called guide wavelength (G), which is dependent on both L and B. G is fairly complex to calculate. We need to know it to evaluate several other factors in the design of the horn, including the position of the probe relative to the closed end of the horn (C), the length of the pickup probe (D), and even the range of total length for the horn (A). Figure 2.9 is a nomogram that will let you quickly determine both G and C (the probe position) for any horn diameter in the permissible range.

You can custom fabricate your own horn from large-diameter tubing stock, but a perfectly suitable horn can be constructed from a 2-pound (908-g) coffee can.

These cans have a diameter of 12.75 cm which is just fine for dimension B, and results in a guide wavelength (G) of 30.5 cm (from Figure 2.9). The useful length of a horn of this type ranges from 0.5G to 1.5G, and the coffee can, with a length of 16 cm, exceeds the minimum length of 0.5G (15.25 cm) and is therefore useful as-is.

The Probe

In theory, the length of the pickup probe is dependent to some degree on its diameter. In practice, as long as the probe ranges from ⅛ inch (3 mm) to ³⁄₁₆ inch (8 mm) in diameter, a length of 0.2L (3.6 cm) is about right. The probe should be positioned about 0.25G from the closed end of the horn; with our coffee can, this distance turns out to be about 7.6 cm.

Horn Construction

The description that follows will use dimensions based on the coffee-can horn. If your horn has different dimensions, use the design equations (Figure 2.8A) and the nomogram (Figure 2.9) to calculate the

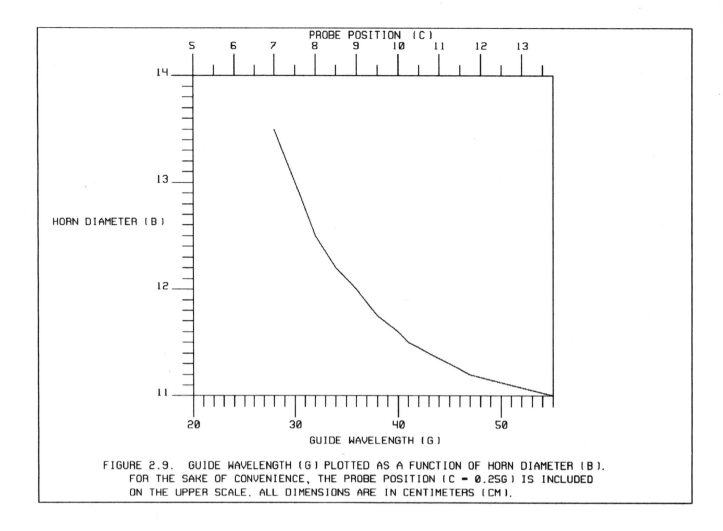

FIGURE 2.9. GUIDE WAVELENGTH (G) PLOTTED AS A FUNCTION OF HORN DIAMETER (B). FOR THE SAKE OF CONVENIENCE, THE PROBE POSITION (C = 0.25G) IS INCLUDED ON THE UPPER SCALE. ALL DIMENSIONS ARE IN CENTIMETERS (CM).

critical dimensions, substituting them in the following description as needed.

Start construction by preparing the probe. The probe is constructed of a 3.6-cm (1.4-inch) piece of 5/32-inch (4-mm) brass tubing available at most hobby shops. Slip the brass tube over the center pin of a Type N female coax connector (UG-58A/U) and solder the tubing in place.

Measure 7.6 cm (3 inches) from the bottom (closed end) of the coffee can and use a 5/8-inch (1.6-cm) chassis punch to make a hole for the N connector/probe. Figure 2.8A shows the N connector mounted with screws, but that will invite problems with a poor ground to the body of the horn, particularly in the event of corrosion. It is best to place the probe through the hole in the wall of the horn and solder the connector into place. To do this, you should place a wire brush in an electric drill and clean the paint from the surface of the can around the hole, heat the bare metal with a propane torch, and tin the area with solder. If the body of the N connector is silver plated, the connector can now be soldered in place using the torch. Some of the new connector plating materials will not take solder. If this is the case with your connector, use the wire brush or a file to scrape off the plating on the mounting flange of the connector to expose the brass underneath. Tin the exposed brass and then solder the connector into place.

We are going to mount four small, brass, right-angle brackets equidistant around the outside of the open end of the horn. Such brackets are commonly available in hardware stores, or they can be fabricated from brass or steel stock. With the mounting areas marked, clean off the paint from each area with a wire brush and solder the brackets into place. Now go back over all the soldered areas with a wire brush to clean off any excess rosin, mask the N connector and probe with tape, and give the entire inside and outside of the can a coat of rustproof primer followed by several coats of a good exterior enamel to protect the can from corrosion.

So far, we have avoided having any unnecessary metal protruding into the horn; that is highly desirable to avoid distorting the delicate balance of RF fields on which the proper operation of the horn depends. We must now mount the horn without having any metal, including screws and other hardware, near the horn opening. All of the RF energy reflected by the dish must pass through the open end of the horn; even small amounts of metal can absorb a significant portion of this energy and needlessly reduce gain. To mount the horn, we need a square of acrylic or Lucite plastic, or unclad G-10 fiberglass board material just slightly larger than the diameter of the horn. We will also need 3 or 4 nylon angle brackets and some no. 4 nylon hardware, all of which should be obtainable from your local hobby shop.

We will mount the plate to a piece of PVC plumbing pipe that will be used as a support to hold the horn at the proper distance from the center of the dish (Figure 2.8B). At your local hardware store or plumbing supply shop, purchase two metal pipe flanges and a short (1-foot) stub of galvanized pipe that will screw into one of the flanges. A 1-inch diameter pipe is about right, but you should choose a pipe diameter that is a slide fit into an available diameter of PVC tubing, so make your final choice of galvanized pipe and matching flanges while checking against the available PVC pipe. Unless you have a very large dish, approximately 8 feet of PVC pipe will be required. While you are at it, purchase a PVC coupler for your pipe size, and a bottle of PVC cement.

Cut off a 12-inch length of the PVC pipe. Take the short piece of PVC pipe and mount 3 or 4 nylon angle brackets flush with one end of the pipe using nylon hardware. Place the horn, face down, on the plastic plate and drill holes to match the mounting brackets around the mouth of the horn. Now, take the plate and, placing the PVC pipe with the nylon brackets at the center of the plastic plate, mark and drill holes to match the nylon brackets. Use nylon hardware to secure the plastic plate to the end of the 12-inch PVC pipe, then mount the horn to the plate using steel hardware. Figure 2.8B shows this overall arrangement. This figure shows the steel brackets bolted to the wall of the horn, but a better solution is to solder them in place. Now use the PVC cement to secure the coupling section at the free end of the pipe and set the assembly aside to dry.

To proceed with final mounting of the horn we need to know the focal length of the dish. Ideally, we want the horn to be positioned so the RF waves from the dish come to a focus just inside the open end of the horn. The focal length (A) can be calculated using the formula:

$$A = \frac{(0.5 \times D)^2}{4 \times Y}$$

where D is the dish diameter and Y is the depth of the dish at its center. To measure Y, lay a board or other straight edge across the face of the dish and measure from the center of the dish to the center of the board edge in contact with the rim of the dish. Either American or metric units can be used in the calculation, but don't mix American and metric units!

Temporarily insert the remaining PVC pipe into the pipe coupling attached to the horn assembly and measure along the PVC pipe to achieve a length equal to the focal length (A) minus 5 inches. If A = 24 inches, for example, you would measure a total length from the horn mounting plate of 19 inches. Remove the pipe from the coupling and cut it at the marked point with a hacksaw. Use the hacksaw to cut several 6-inch long slits in the far end of the PVC pipe. Now, use the PVC cement to insert the uncut end of the pipe into the coupling on the horn assembly and let the joint cure. Bolt the two pipe flanges, back to back, on both sides of the center of the dish and install the pipe stub (galvanized) so that it protrudes from the front center of the dish. Slide a stainless-steel hose clamp over the slit end of the PVC pipe, and slide the PVC over the galvanized pipe at the center of the dish. Twist the PVC horn mast so the probe is vertical, relative to the ground, and tighten the clamp so that the end of the PVC pipe mast is 3 inches from the dish surface.

If the PVC mast supporting the horn is not too long, guying of the horn may not be required. Simply connect a length of low-loss RG-8 coax (with a type N plug) to the connector on the horn, and run the line to the edge of the dish, or tape it along the length of the PVC mast. If the mast is quite long, or you'll be mounting a preamp right at the horn feed point, guy the horn using nylon fishing line from the corners of the horn mounting plate to the periphery of the dish.

Final Setup

Once your equipment is ready, use the information in Chapter 8 to point the antenna as accurately as possible in both azimuth and elevation. Listen for the satellite signal when the transmission schedule indicates the desired satellite to be active, and make minor adjustments in both azimuth and elevation to optimize the signal strength. Once you have achieved the best signal with antenna adjustments, loosen the clamp at the bottom of the PVC mast and slide the mast in and out on the pipe stub to further improve the signal. This latter step will optimize the position of the horn relative to the dish focal point. Prior to tightening the clamp, rotate the mast in either direction to optimize the polarization. When no further improvement is obtained, tighten the clamp and go inside to enjoy your new WEFAX installation!

COAXIAL TRANSMISSION LINES AND CONNECTORS

It is a fact of life that we must move RF signals around between antennas and various pieces of equipment; for that, we need coaxial cable transmission lines and the matching connectors. Losses are inevitable in RF cables and we need to minimize such losses whenever possible. Losses are a function of several factors including frequency, cable diameter, braid density, dielectric material (the insulating material—usually white—between the braid and the center conductor), and the diameter of the cable.

Operating Frequency

In a given type and grade of cable, losses will rise dramatically with operating frequency. This means that S-band losses will be considerably higher than comparable losses in the 137-MHz VHF range. This translates to the need for shorter runs of cable and the use of higher-quality cable for S-band.

Braid

The conductive braid beneath the outer plastic jacket of the cable is critical. Ideally this sheath should be solid metal and, indeed, the lowest-loss cables have such solid jackets, but the cables are expensive and inflexible. When the jacket is braided fine wire, as it is with most cable, braid density becomes critical. Bargain cables will have a low braid density: in some, you can even see the dielectric through the braid! A good cable will have very dense braid, and the fine copper wires that make it up will be plated to reduce corrosion. More-expensive cables will have a double thickness of braid, but we do not usually have to go to that extreme if we are careful in other aspects of cable selection.

Dielectric

The most commonly available cables have a dielectric of solid polyethylene. Still-better cables will have a foam polyethylene dielectric; these exhibit lower losses. Such cables are often designated as "foam," or carry an F in place of the normal R in the cable designation. Teflon—and even air—can be used as a dielectric to achieve even lower losses, but such cables tend to be very expensive.

Cable Diameter

All other factors being equal, losses are higher in small-diameter cables. The two generally available cables for 52 ohms impedance (the value we will use throughout most of the station) are type 8 (0.405 inch diameter) and type 58 (0.195 inch). The standard designation for the 8 cable is RG-8 with the foam types sometimes labeled F-8. The type 58 cables are represented by RG-58 with F-58 occasionally used for the foam version. Type 58 cable is suitable for short runs between equipment at VHF, but for any cable runs over 50 feet long you should use type 8. A good-quality type 8 foam will support runs of up to 10 feet at S-band,

and you should not use type 58 at these frequencies for any runs over a few inches!

A great many manufacturers make coaxial cables, and each makes a number of grades. Although price is not an infallible guide, a higher price usually equates with better materials, quality control and lower losses. Avoid bargain CB-type cables as they are likely to exhibit high losses even at VHF and are totally unsuited for S-band use. In the balance between cost and quality, I recommend Belden 8214 for your type 8 cable needs, and Belden 8420 for type 58. In general, try to obtain cables of at least this quality to avoid unanticipated problems with your system.

Connectors

Like cables, connectors can generate both losses and instability. Constant-impedance connectors do not disrupt the impedance of a given line and will improve performance over connectors that are not matched, creating an impedance "bump" in the line with every connector set!

The most-readily available connectors are the so-called UHF series. The basic plug (male cable mounting connector) is the PL-259 designed for direct use with type 8 cable, and with type 58 and 59 (70-ohm) cable with the use of screw-in adapter sleeves. The matching chassis-mounting female connector is the SO-239. Threaded barrel connectors are available for connecting two PL-259 equipped cables. These connectors are not constant impedance types, and are marginally useful at VHF and completely unsuited for S-band.

A much better connector for VHF use, particularly with type 58 cables, is the BNC series. The military designation for the cable mounting plug is UG-88U. The matching flange mounted female connector, used for preamp and receiver inputs, is the UG-447/U. The UG-625/U is similar except that it uses thread mounting to the chassis wall. This is the connector of choice for your VHF connections using type 58 cables. These connectors can be used at S-band in a pinch, but regular use should be avoided.

The premier VHF connector and the most affordable option for S-band is the Type N series of connectors:

UG-21B/U	Type N plug for 8-series cable
UG-536B/U	Type N plug for 58-series cables
UG-58A/U	Type N flange mounting connector (receiver and preamp inputs)
UG-30/U	Type N double female connector (bulkhead mounting, weatherproof enclosures, etc).

No matter which connectors you employ, proper installation is a must for the best service and weatherproof integrity in the case of the N series cables. Manufacturers' data sheets cover proper installation as do publications such as *The ARRL Antenna Book* and *The ARRL Handbook*. All outdoor connections should be wrapped with plastic electrical tape, sealed with the putty-like sealer available at most outlets selling TV or other antenna components, and wrapped again with plastic electrical tape. Corrosion of connectors and moisture in transmission lines are the two major enemies of a trouble-free antenna system!

OPTIONS

If you have the misfortune of living in a situation that does not permit you to erect a permanent antenna system, you can still enjoy this hobby with the exercise of a little ingenuity. Let's assume for a moment that you live in a subdivision that does not permit antenna systems or you rent from a landlord that will not permit you to install an antenna. One solution would be to use the VHF Yagi, mounted on a short pole to make it easy to handle. With coax running back into the house, you can simply track the satellite manually from the backyard while a simple digital timer is used to activate the station recorder to tape the pass while you are working the antenna. With a digital watch and your tracking data on a 3 × 5 card, you can log the entire pass with no fuss, taking the antenna back inside when it is not in use. The neighbors may think you are a bit odd, but the technique works just as well as the most-elaborate antenna system. With the addition of a pair of headphones you can even track Meteor satellites with the simple linear Yagi, simply rotating the antenna to maximize the signal as you track the satellite!

Similar approaches can be used with a collapsible S-band dish, and it will even work when used from an apartment balcony for some satellites and/or passes.

If you live in an urban apartment with no balcony you still have options. The key here is to make your receiving system completely portable. A typical receiver will operate from a 12-V dc battery pack, and battery-operated stereo recorders are available. You will need to duplicate the 2048-Hz clock circuit from the scan converter in Chapter 6 so you have a reference tone for the left channel, but those circuits will work from a battery supply with no problems. If you have access to the roof of your building, you can track and record a pass with no difficulty. Even if the roof is off limits, a neighborhood park or the parking lot of a shopping center can supply a site with enough open sky for effective tracking. With enough ingenuity, you can enjoy this hobby in almost any environment. It may not be as convenient as having everything permanently installed, but you will be able to enjoy the pictures that you do receive. If you have the interest, don't let a restrictive situation inhibit your creativity—it can be done!

Chapter 3

Weather-Satellite Receivers

INTRODUCTION TO VHF RECEIVERS

The most important link to that vehicle in space is your station receiver. You cannot compromise here. For, unless the receiver meets certain minimum requirements, you may be able to hear the satellite with a strong and consistent signal, but you'll *not* be able to obtain a satisfactory picture. To understand this seeming paradox, it's necessary to understand some aspects of the RF formats employed in weather-satellite transmissions of potential interest to amateurs.

All of the weather-satellite transmissions in which you are likely to be interested are transmitted as FM signals, and all polar-orbiter transmissions are made in the 137- to 138-MHz frequency range. Our basic FM receiver is made up of a series of circuit modules along the lines of the block diagram shown in Figure 3.1. In order to make a rational choice in terms of receiver selection, we need to know a bit about how each of these circuit elements functions. If you are interested in delving in depth into receiver design, chapters in a current edition of the *ARRL Handbook* provide comprehensive coverage. The following discussion, while brief, touches on some of the major factors that must be taken into consideration.

RF Preamplifier/Mixer

The VHF signals we receive from the weather satellites are quite weak. That shouldn't be surprising once you know that the satellite transmitter is rated at about 5 watts output (enough power to light a flashlight bulb), and the satellite's antenna has little gain. The job of the receiver RF amplifier and mixer stages is to provide some gain and selectivity before the signal is converted to a lower frequency for additional amplification and detection.

The gain and noise-figure characteristics of the receiver front end are two of the primary factors that determine how effective the receiver is. Because the satellite signal is quite weak, you might think that the gain of the RF amplifier stages (how much they amplify the incoming signal) is the primary concern, but this is not the case. Gain is secondary to front-end noise figure.

All electronic devices—including the first RF amplifier stage of our receiver—generate random noise. The noise generated by the first amplifier stage will be amplified (along with the desired signal) by all later stages in the receiver. If the first amplifier stage generates a lot of noise, we may never be able to hear the satellite signal, no matter how much additional amplification the receiver provides. The first amplifier stage needs to provide some amplification to overcome losses in the first mixer stage, but it must do so while adding the smallest amount of random noise to the signal—hence, our worship of low-noise front-end circuits.

In the late 60s and early 70s, typical tube-type amplifiers could provide front-end noise figures of 3 dB or more. Very expensive tubes (do you remember the venerable 417A?) might push this figure down to around 2 dB, but circuit design and adjustment were tricky. Early bipolar transistor amplifiers could perform as well as some of the better tubes, but were easily overloaded by strong signals. The introduction of *field effect transistors* (FETs) was a major advance. *metal oxide* FETs (MOSFETs) can provide a 2.5-dB noise figure with no fuss, and specially designed *junction* FETs (JFETs) are capable of 2-dB noise figures with little potential for overloading. FET technology has advanced steadily and the latest designs, fabricated using *gallium arsenide* technology (GaAsFETs) easily achieve noise figures below 1 dB!

Although an overall reduction from 3 dB to 1 dB may seem small, remember that the decibel scale is logarithmic. An amplifier with a 3-dB noise figure limits the weak-signal performance of the receiver. The same amplifier, with a noise figure of less than 1 dB, is limited by external noise sources, either man-made or by the natural RF emission of the sun or other stars in our galaxy or beyond! Therefore, GaAsFETs are the devices of choice for the first RF amplifier stage. Succeeding amplifier stages (a receiver rarely has more than two stages of RF amplification) and the mixer also contribute noise, but the effect of this noise

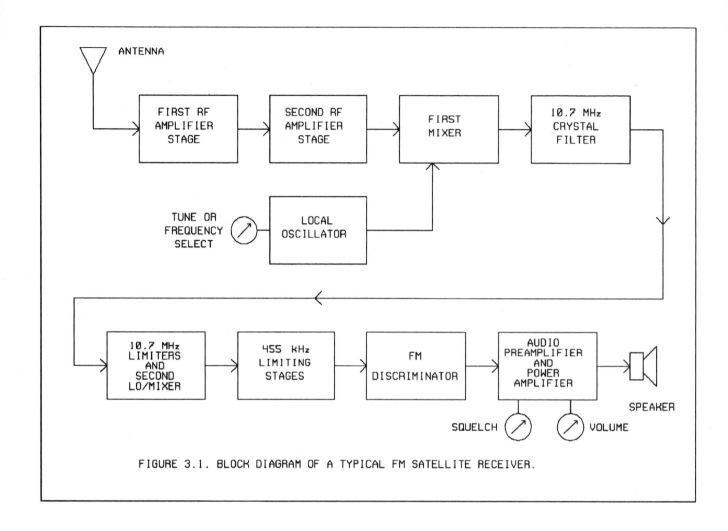

FIGURE 3.1. BLOCK DIAGRAM OF A TYPICAL FM SATELLITE RECEIVER.

is negligible provided the first stage has sufficient gain to—in effect—set the system noise figure.

Note, too, that signal losses in transmission lines prior to the first RF stage have the effect of *adding* to the effective noise figure of the front end. If we have a 1-dB noise-figure RF stage fed by a transmission line with 3 dB of signal loss, our effective front-end noise figure is 4 dB! Many of the newer receivers boast the use of a GaAsFET front end, but we cannot realize the full potential of the device if we precede it with excessive transmission-line losses. If we place a low-noise GaAsFET stage *at the antenna*, we can overcome the problem of transmission-line loss entirely.

The total gain of the RF front end should only be that required to effectively set the system noise figure and overcome any losses in the first mixer stage. Too much gain can overload the mixer, creating spurious products that degrade receiver performance. Excessive gain can also lead to the potential for overload and receiver desensing from unwanted signals outside of the desired frequency range. This is a major problem in most metropolitan areas due to the concentration of all sorts of RF-emitting devices. If you have a very

hot external preamplifier and relatively low transmission-line losses, you may actually have to reduce the gain of the receiver first-amplifier stage to avoid the overload problem. If gain is not easily altered, you may find it worthwhile to introduce some additional cable losses, or use an input attenuator to keep signal levels below the overload threshold.

Local Oscillator/Mixer

The mixer stage that follows the first RF amplifier(s) is actually a frequency converter. It mixes (or combines) the incoming signal with an RF signal from the local oscillator (LO) to produce outputs at several frequencies, including the LO frequency (L), the signal frequency (S), and two additional frequencies, S + L and S – L. In most basic receiver designs, we want to convert the signal to a lower frequency for effective amplification and filtering: an *intermediate frequency*, or IF. The S, L, and S + L frequency components are suppressed with tuned circuits that are adjusted to pass the S – L component with minimal losses. In simple receiver designs, the first IF is typically at 10.7 MHz.

If we want to convert a 137.5-MHz signal to 10.7 MHz, we can accomplish the job with either one of two LO frequencies: 126.8 MHz (137.5 − 126.8 = 10.7), or 148.2 MHz (148.2 − 137.5 = 10.7). An LO that is 10.7 MHz below the signal frequency (*low-side* LO injection) is usually chosen for circuit simplicity, but there is no reason why *high-side* injection (L = 10.7 MHz above S) cannot be used.

The LO signal thus determines what frequency is processed by the receiver. There are three basic ways to generate the LO signal:

- Use a tunable or variable-frequency oscillator (VFO)
- Use a crystal oscillator
- Generate the signal by digital synthesis

The VFO approach can provide continuous tuning over a specified frequency range, but at VHF, it is difficult to design a VFO that will be stable enough to provide reproducible calibration of the tuning range. A crystal oscillator can provide the needed stability, but requires the use of a different crystal for each frequency (or channel) we want to receive. The most advanced approach is to synthesize the LO signal digitally, providing the ability to tune a given frequency range in discrete steps, depending upon the design of the synthesizer. Most of the synthesizer circuits we'll encounter can provide tuning in 5-kHz steps.

IF Circuits

The receiver IF stages serve two roles. The first is to provide gain: the amplification of the signal prior to detection. In modern receivers, the IF stages consist of integrated circuits (ICs) that can provide very high gain (often greater than 100 dB) in the form of a simple chip. Most basic receivers actually have two blocks of IF circuits. The *first IF* is typically at 10.7 MHz. That is followed by a second mixer and crystal LO (often integrated into the 10.7-MHz IF-amplifier chip) that converts the signal to a 455-kHz *second IF* that provides still more signal gain. Very high gain in the IF stages is necessary because FM detection requires that we eliminate all of the amplitude variations in the incoming signal. The high-gain IF circuits accomplish this with *limiting* amplifier circuits, where all signals within a wide range will have a constant amplitude by the time the signals reach the FM detector circuits.

A second important role for the IF is to provide filtering of the signal. The IF stages can be looked upon as providing a frequency window, passing signals within a certain frequency range or *bandwidth* while providing significant rejection of signals outside of that range. The reasons for IF filtering are two-fold: to reject other unwanted signals that fall near the 10.7-MHz (or 455-kHz) IF, and to minimize the effects of noise.

The information in an FM signal is carried in the form of deviations from the basic signal frequency—hence the name *frequency modulation*. If the IF bandwidth is too narrow to accommodate the deviation (frequency swing) of the signal, some of the signal is clipped off by the IF filters and the signal will become distorted and generally unusable. On the other hand, if the IF is too broad, it will pass unwanted signals and noise that degrade the desired signal. Ideally, we want the IF to be wide enough to pass the signal with full deviation, with some additional allowance for errors in LO injection and changes in the signal frequency caused by Doppler shift.

An ideal satellite receiver has an IF bandwidth of about 40 kHz. The bandwidth is typically set by crystal or ceramic filters at the input of the 10.7-MHz IF stages, and by ceramic filters or simple tuned circuits at the 455-kHz stages.

Detector/Audio Stages

There are many ways to convert the varying signal frequency to a varying voltage that can be amplified to drive a speaker or other devices. Generally, the circuits that perform this job are known as *discriminators* or *ratio detectors*. In modern receivers, there is a trend toward the use of phase-locked loop (PLL) discriminators because of the excellent linearity typical of such circuits. The typical PLL discriminator is followed by a low-noise audio preamplifier and a power amplifier (usually a single IC) stage to drive a speaker.

Not shown in Figure 3.1, but present in most receiver designs, is provision for squelch control of the audio output. When no signal is present, the noise output of the receiver can be very distracting. The job of the squelch circuits is to shut off the audio output when no signal is present, and to turn on the audio stages when a signal appears. Squelch circuits can operate by detecting the presence of the signal either directly or indirectly by means of decreasing noise when a signal is present. The threshold point for squelch operation is usually adjustable by means of a squelch control.

Frequency Coverage

Our frequency range of interest is fairly narrow, with all polar-orbiter transmissions confined to the 137- to 138-MHz range. US TIROS/NOAA polar-orbiter transmissions are made on frequencies of 137.50 and 137.62 MHz. Each of the two operational satellites that are ideally in service at any one time use one of these frequencies. NOAA-10, for example, uses 137.50, while NOAA-11 uses 137.62. Each satellite has backup capability on the alternate frequency should a primary transmitter failure occur. Soviet

Meteor/COSMOS weather-satellite transmissions occur on a wider variety of frequencies. In the past, 137.15 and 137.30 MHz were the two prime frequencies and there still appears to be an operational Meteor satellite, and often more than one, on 137.30 MHz most of the time. In recent years, 137.85 MHz has been used quite regularly, and 137.40 MHz has recently been placed into service as I prepare this edition. In any case, your receiver must cover 137.50 and 137.62 MHz; 137.30 and 137.85 MHz are desirable options.

Reception of geostationary WEFAX transmissions on 1691 and 1694.5 MHz (the latter used only by the European METEOSAT) is usually accomplished by using a converter ahead of the basic VHF FM satellite receiver. Such converters are designed so that the desired signal comes out at one of the "standard" VHF satellite frequencies (usually 137.50 MHz).

Receiver Bandwidth Considerations

Receiver bandwidth turns out to be one of the biggest hurdles to overcome since all of the various satellites use deviation values that are significantly higher than those employed for standard FM voice links. The biggest market for FM receivers (if we omit FM broadcast and TV sound) is for various kinds of scanners operating in the police and public-service bands. These transmissions typically deviate a maximum of ±7.5 kHz, and the receivers usually have a 15-kHz IF bandwidth. Unfortunately, if you tally up the values for signal deviation for a satellite such as the TIROS/NOAA series (±18 kHz) and Doppler shift (±3 kHz), you end up with a required bandwidth in excess of 40 kHz!

A signal from one of the polar orbiters, received on a typical scanner with 15-kHz bandwidth, will be severely distorted and won't produce a usable signal, though an unmodulated carrier sounds fine. Although deviation levels of the Soviet Meteor polar orbiters and the geostationary GOES and METEOSAT satellite are lower than those of the TIROS/NOAA satellites, their signals are still too wide for satisfactory reception with a stock 15-kHz-bandwidth receiver.

Receiver bandwidth is typically set by one or more filters in the IF chain. There is almost always a crystal or ceramic filter at the 10.7 MHz first IF, and there is often another filter (usually ceramic) at the 455-kHz second IF. If the receiver employs a crystal or ceramic filter at 10.7 MHz, the IF bandwidth can often be widened by simply replacing the original filter with an inexpensive 30-kHz crystal filter. A cheap 30-kHz filter has very wide skirts, and you'll end up with a receiver with sufficient bandwidth to do the job. In contrast, a good 30-kHz filter has steep skirts, and the receiver will still be too sharp for TIROS/NOAA service. If the receiver also uses a 455-kHz ceramic filter, you may

have a problem finding a replacement for it because wider-bandwidth ceramic filters at this IF are harder to come by. I have had good success in some instances by simply replacing the 455-kHz filter with a small-value (5- to 15-pF) coupling capacitor.

VHF RECEIVER OPTIONS

When selecting a receiver, new or used, its IF bandwidth should be your first consideration. In the sections that follow, I'll outline some general strategies for receiver selection and modification.

Commercial Satellite Receivers

Obviously, the simplest approach is to simply buy a receiver designed for satellite reception. Vanguard Labs has been one of the principle vendors for affordable satellite receivers over the years. Their present receiver line contains a new version of an old standby, plus a brand new product offering.

Figure 3.2—The Vanguard Labs Model FMR-250-11, an 11-channel, crystal-controlled VHF FM receiver. Many are still in service, but it is no longer available from Vanguard.

The FMR-250 (Figure 3.2) is an 11-channel, crystal-controlled receiver featuring dual conversion, a GaAs-FET front end, and a PLL FM detector. (Vanguard has replaced this receiver with its WEPIX 2000.) Volume, squelch, and frequency selector switches are on the front panel. The receiver has an internal speaker as well as a jack for hooking up an external speaker or audio network. The receiver comes supplied with a single crystal and operates from a 12- to 14-V dc supply. The latest receiver version comes complete with a wall-mount dc supply. (But you may want to use another supply; see **Receiver Modifications**.) A toggle switch mounted on the rear panel permits you to apply or bypass 12 V dc to the antenna transmission line to power remote preamps or converters. Although earlier models of this receiver had a 30-kHz-wide IF filter, the filter was too sharp for optimum polar-orbiter reception. The latest model features a slightly broader IF filter for optimum performance.

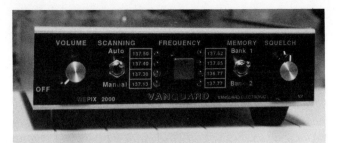

Figure 3.3—The new Vanguard WEPIX 2000 synthesized receiver is housed in an enclosure the same size as that of the FMR-250 series, but the front panel includes a number of new displays and controls associated with the internal frequency synthesizer. Eight LED indicators mark the frequencies programmed by the manufacturer (137.13, 137.30, 137.40, 137.50, 137.62, 137.85, 136.77 and 137.77 MHz). A scanning mode toggle switch enables automatic sequential scanning of all channels, or can be set to provide manual stepping through the channels using the square push-button switch in the center of the panel. A memory bank toggle switch selects either Bank 1 (the frequencies noted above) or Bank 2 (eight channels at 5-kHz intervals, centered on 137.50 MHz). A 5-pin DIN socket on the rear apron carries the audio signal and TTL lines.

The WEPIX 2000 receiver (Figure 3.3) represents Vanguard's latest offering, and combines a number of features that bring this receiver very close to being optimum. A total of 16 frequencies, in two banks of 8, are programmed into the unit using an *e*rasable *pro*grammable *read-only m*emory (EPROM) chip. The first bank of 8 channels includes all active polar-orbiter frequencies, including those for the planned Chinese polar-orbiter satellite now scheduled for launch sometime this year. The second bank of 8 channels represent a series of channels, separated by 5 kHz, centered on 137.5 MHz. This bank is designed for use with a WEFAX downconverter, permitting fine-tuning of the IF to compensate for frequency errors due to thermal drift or other factors that would shift the output frequency of the downconverter. The active frequency is indicated by an associated LED. You can step through the channels manually, or use the receiver as an automatic scanner. Complete information on programming EPROMs for other frequencies is included. Up to 250 channels can be programmed into the EPROM; additional bit space is available for future control functions.

The receiver uses GaAsFETs for both the front end and mixer stages, assuring a low front-end noise figure. The receiver is a double-conversion design with a crystal IF filter and noise-operated squelch. A variable-gain output is available for driving your display system or recorder along with an independently adjustable speaker volume control. Two TTL outputs (one normally high and the other normally low) are provided for control of a tape deck or other external equipment. The receiver is priced at $330, and is ready for use right out of the box.

Kits

Your most economical option is to build a receiver from a kit. The number of VHF FM receiver kits available is now quite limited, but fortunately Hamtronics still has several models, one of which is specifically broadband enough for satellite use (the RX75A). Such receivers are typically crystal controlled.

Converter/Communications Receiver

Another possibility, not often considered, is to use a converter ahead of a communications receiver equipped with an optional FM IF. If the FM option provides a sufficiently wide bandwidth, or can be modified to do so, this approach can be very effective. Older, HF-receiver VFOs didn't have a high degree of resetability, so it was difficult to return to and monitor a specific frequency. That problem doesn't exist with most modern HF receivers. You can modify a standard 2-meter converter to work into a 10-meter IF (add a new LO crystal and retune), or you can contact companies like Spectrum International and have them modify a converter for you. Recently, Hamtronics developed a converter that handles satellite-signal frequencies at the input and has an output in the 28- to 30-MHz range. Again, double-check the FM IF bandwidth specs: If the receiver IF bandwidth is too narrow, you'll get a strong but unusable signal!

Scanners

Now we come to the subject of scanner receivers! These radios come in three general configurations:

- First-generation crystal-controlled units.
- Programmable units covering the public-service bands.
- Wide-coverage programmable units.

Before discussing each category, a few general guidelines are in order. First, if you are going to buy a new unit, insist on seeing a schematic of the radio. In the case of the Radio Shack receivers, such a schematic is usually in the manual (although you may need a magnifying glass to read it). The purpose of this step is to see where the filters are located in the circuit. If you buy a used scanner, try to obtain a Sams Photofact sheet on it from your local electronics distributor. Try to stick with units that employ standard IFs (10.7 MHz and 455 kHz) as you'll have more luck in locating replacement filters to widen the IF bandwidth. (I per-

Figure 3.4—An example of one of the newer wide-range programmable scanning receivers, the Regency MX-5000. Although such receivers provide the necessary frequency coverage, their wide IF bandwidth presents a real problem. Also, the very wide tuning range of these receivers makes them susceptible to undesired signals.

sonally avoid receivers that use IFs other than those.) See if you can get the dealer to open the radio's case for you. (Most won't do so, but it's worth a try!) A peek inside the radio gives you a chance to assess the mechanical aspects of filter replacement.

Crystal-controlled scanners are by far the simplest to work with. Generally, the board layouts are more open, making it easier to get at the filters, and the VHF HI band (144-174 MHz) can usually be retuned to cover 137 MHz. Don't expect spectacular sensitivity, because the front ends are designed for broadband service. On the other hand, don't worry about it either: An external preamp will be needed anyway.

Programmable scanners designed strictly for public service use present two areas of concern. The first involves the tuning range. Most programmable scanners are strictly limited to 144- to 174-MHz coverage in the VHF HI range; entering a 137-MHz satellite frequency will only get you an error message. Two solutions to this problem exist: Set the receiver to a frequency that provides high-side LO injection for a satellite frequency, and retune the RF stages to peak at 137 MHz. Or, use a converter to move the 137- to 138-MHz satellite band coverage to a frequency range in the VHF LO band (usually 30 to 50 MHz). Hamtronics manufactures a converter (the CA144-6) to do the latter job.

The second problem concerns IF bandwidth. The PC boards of these receivers tend to have tight component layouts. Getting at and replacing filters can be a

problem. The degree of miniaturization makes it imperative to locate equally-small replacement filters; this can be difficult. Here is where examining the circuit and looking inside the radio can pay dividends prior to purchase!

The most impressive scanners of all are the wide-coverage units that represent the new generation of scanning receivers. A unit such as the Regency MX5000 (Figure 3.4), for example, tunes from 25 to 550 MHz in 5-kHz steps. The radio features programming of AM, narrowband FM, and wideband FM for any channel; it's equipped with 20 memory channels and has all sorts of fancy search options.

At first, a receiver of this sort looks ideal because it covers the required frequency range, and the wideband FM mode (designed for TV sound and FM broadcast reception) is certainly wide enough for the satellite signal. Unfortunately, the wideband mode is *very* wide, resulting in a severe degradation of the signal-to-noise ratio.

A GaAsFET preamp at the VHF antenna will usually minimize noise and maximize signal to the point where a receiver such as the MX5000 is useful for polar-orbiter service, but its use for reception of geostationary WEFAX signals presents another problem. Even a GaAsFET preamp for 1691 MHz is significantly noisier than one used at 137 MHz, so the only solution for boosting the signal without introducing more receiver noise is to have higher antenna gain. You'll have relatively poor results in trying to use a small (4-ft [1.2-m]) dish with such a receiver, even though it delivers a satisfactory signal when used with a receiver having a 30- to 40-kHz IF bandwidth. The system will work if you use a GaAsFET preamp and a larger antenna (10 ft [3 m]) such as a TVRO dish, but all of this adds to the expense of your system, not to mention the required antenna space. All such problems would disappear if the narrow-band FM filters could be replaced with 30- to 40-kHz units, but the circuit board of a receiver like the MX5000 is packed to the limit with the smallest components you'll see outside of an IC. (I have no intention of attempting surgery on mine, and I am usually quite casual about such modifications!)

Major receiver manufacturers such as ICOM, Kenwood, and Yaesu offer several synthesized receiver models that tune all of our VHF frequencies; some even tune the WEFAX S-band frequencies as well. Most of these receivers are in the $500 to $1000 price class. Although they're engineering marvels, they have the same IF-bandwidth limitations as the less-expensive programmable scanners. They are much larger and more impressive-looking than the simpler receivers noted earlier and may be cost-effective if you have other VHF and UHF monitoring interests beyond satellites. Dealers for these receiver lines can be found in the advertising pages of Amateur Radio magazines such as *QST, 73 Magazine* and *CQ,* as well as more specialized publications such as *Popular Communications* or *Monitoring Times.* A good public or university library will have one or more of these magazines. Discount pricing can usually be obtained. If weather satellites are your primary interest, one of the commercial satellite receivers, matched for the proper IF bandwidth and optimized for our frequencies of interest, is a far better buy.

Surplus Receivers

In years past, using receivers retired from commercial FM service (police, fire, and other public-service uses) was a violable option. Older tube units featured 30-kHz IF bandwidths and are suitable for use with suitable frequency modification and retuning. Unfortunately, parts for such receivers are hard to obtain, and the use of an outboard preamplifier is a necessity to get a reasonable noise figure. Newer surplus units are solid state, but typically are designed for 15-kHz IF bandwidths, requiring additional modifications. Such receivers are of extremely high quality, however, and if you can obtain one inexpensively and feel confident in working with the conversion, the approach can be worthwhile.

Receiver Modifications

Regardless of which approach you take to acquiring the station receiver, a few additional modifications will be useful. The first concerns power supplies. A *very* well-regulated 12-V supply should be used. Ac-operated scanner supplies almost always have an unacceptable hum level. You may not notice the hum on voice reception, but it *will* show up in satellite pictures. Fortunately, most receivers have provisions for using external 12- to 14-V power supplies.

A second desirable modification is to provide a constant-amplitude audio tap from the receiver to your display system and tape recorder. This makes the video drive level independent of the receiver audio gain control and is quite handy. The easiest way to accomplish this is simply to tap off the top of the volume control (point furthest from ground) using a 0.1-µF disc or Mylar capacitor, and bring the line out to a phono jack at the rear of the receiver.

Preamplifiers

The use of an external preamplifier, preferably located at the antenna to reduce the effect of line losses, is highly desirable in any case, and a must in the case of the Zapper antenna described in Chapter 2. You'll also need a preamp if you plan to use the wideband FM option for the new generation of programmable scanners. Several vendors, including Spectrum International, Vanguard Labs, and Hamtronics (Figure 3.5) have suitable preamplifiers with some models

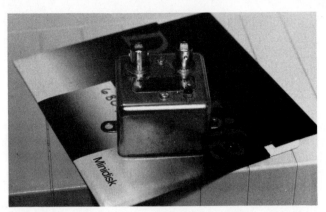

Figure 3.5—An external GaAsFET preamplifier provides the most effective single step in converting almost any receiver to state-of-the-art performance. This Hamtronics unit costs approximately $40. Locating the preamplifier as close to the antenna as possible helps overcome the problem of transmission-line losses.

equipped with weatherproof housings. Generally, these units are of two distinct types: JFET and GaAsFET models. JFETs are very rugged and can provide noise figures of about 2 dB. The new generation of GaAsFETs can provide a noise floor below 1 dB. These are more delicate devices, however, that should not be exposed to intense RF fields.

Decisions! Decisions!

Given all the options for selecting a receiver, is there a best choice? Despite the diversity in the marketplace, if you are acquiring a receiver strictly for satellite use, the options are quite clear. One of the commercial satellite receivers is really your best buy. Any of these units has an almost unlimited service life. Repairs, if they are ever required, won't be costly. Considering the overall importance of the receiver, doing the job right should be your first objective. A crystal-controlled model is somewhat less expensive than a synthesized receiver, but only if you don't consider the final cost of all the crystals you might eventually add!

If finances are a problem, one of the Hamtronics kits will do a fine job, with you providing "sweat equity" (your labor for the construction of the unit). Remember that you'll have to add to the purchase price of the kit the cost of a cabinet, speaker, controls, and any crystals. Older model crystal-controlled scanners have very little resale value and it is often possible to find one for almost no cost. Such a unit will provide useful service with replacement of the IF filters and the addition of a modern GaAsFET preamplifier.

A wide-ranging scanner costs as much—or more—than a commercial receiver such as one of the Vanguard units, and won't perform as well. The more sophisticated receivers will easily double or triple your costs with no increase in performance. Unless you already have one, purchasing such a receiver is only justified if you have other VHF/UHF monitoring interests that require the added frequency coverage.

Trying It Out

Preliminary checks on the receiver can be made with a quarter-wave-vertical (20-in [51-cm]) antenna connected directly to the antenna jack. With such a setup, you should be able to clearly hear both TIROS/NOAA and Soviet Meteor satellites during the best parts of a good pass, provided the receiver is outside (or in a wooden-frame building). There will probably be considerable fading and noise, but you *will* hear the satellite. Once the receiver is known to be working, it can be connected to the operational antenna/preamplifier system.

WEFAX DOWNCONVERTERS

WEFAX transmissions from the GOES, METEOSAT, and GMS satellite are made in the low-microwave region (S-band) at frequencies of 1691 and 1694.5 MHz. The most practical approach to receiving such signals is to use an S-band antenna in conjunction with a downconverter feeding the VHF FM station receiver. The job of the downconverter is to amplify the signal from the antenna, convert the 1690-MHz signals to the 136- to 138-MHz range, provide some VHF amplification to overcome mixer losses and have enough gain to eliminate the losses in the feed line between the converter and the VHF receiver.

When the Second Edition of the *Weather Satellite Handbook* went to press, amateurs were at the forefront of developing suitable converter options, despite the fact that early NASA publications on the new GOES system suggested that small ground stations were probably not practical. Now, we are in the position of having numerous off-the-shelf converters and receivers ready to do the job. Most amateurs today won't choose to embark on the relatively technical exercise of constructing their own converter. It's still important, however, to understand some of the major technical aspects of downconverter performance, if only to make an intelligent choice as to which system to purchase.

GOES WEFAX Link Analysis

When aerospace engineers plan a space communications link, they don't guess what transmitter power levels, antenna gain, and receiver sensitivity will be required to accomplish the desired communications. They engage in a very precise mathematical exercise known as a *link analysis* to predict the performance of the system. When we contemplate setting up a GOES receiving system, it's highly desirable to go through the same steps to determine whether we can make our system work with the equipment on hand, or within the limits of the funds that can be allocated to the project.

Aside from the outright purchase of a computer system for your station, WEFAX ground-station components will represent your single biggest investment. If you're going to plan and spend wisely, you need to take a very analytical approach to receiving system design. If you don't do so, you may find yourself with a relatively expensive collection of hardware that doesn't deliver the performance you require. If you prefer, you can skip the math and get to the bottom-line elements of system analysis. It's helpful, however, to know why you have to trade off on various elements of your system, so you might try to follow the basic elements of the math, even if it seems a little intimidating at first.

The basis of the link analysis is the comparison (under various conditions) of the power level of the satellite signal to the power level of the noise generated by your receiving system. The basic unit of power is the watt (W), but the power levels we'll be dealing with are so small, that it's more convenient to reference power levels to a much smaller unit—the milliwatt (mW), equal to 1/1000 of a watt. Our link analysis has three primary steps: the calculation of the power level of the satellite signal on the ground, the calculation of receiver noise power under various conditions, and, finally, how much antenna gain is required to deliver a useful signal with various receiver systems.

Satellite Ground-Signal Level

Path Loss

The satellite's signal power on the earth's surface is a function of two primary factors—the power of the signal radiated by the satellite and the *path loss* as the signal travels from the satellite to the ground station. The power radiated by the satellite—the *effective radiated power* (ERP) is a function of many factors, including the output of the transmitter, coupling and transmission-line losses between the transmitter and the antenna, and the gain of the antenna. Fortunately, we can bypass all these complexities by simply looking at the manufacturing specifications that indicate an ERP of +54.4 dBm (+54.4 dB relative to 1 milliwatt), or approximately 260 W. The path loss depends on two factors: the operating frequency and the length of the path. Although the frequency is a constant (1691 MHz), the path length is variable, depending upon the look angle between the ground station and the satellite (see Chapter 8). A reasonable "average" value is 188 dB. Since this is a *loss*, we can compute the ground signal level by subtracting the path loss from the ERP of the satellite:

+54.4 dBm (ERP) −188 dB (path loss) = −134.4 dBm

The power available from the satellite transmitter is obviously extremely minute, but to know what it means in practice depends on the relative noise power of our receiving system.

Receiver Noise Power

The noise power level at the input of our receiving system is critical. All electronic devices generate noise and, unless the level of power available from the signal exceeds the internal-noise power level by some reasonable margin, we won't receive a usable signal. The receiver noise power (TNL) can be calculated if we know two factors—the receiving system noise figure (NF_{dB}) and the system bandwidth (BW_{Hz}):

$$TNL_{dBm} = -174 + 10(\log(BW_{Hz})) + NF_{dB}$$

Because we want TNL to be as low as possible (so we can hear weak signals), we need a low noise figure, because noise figure *adds* to the noise power. Bandwidth is significant: The wider the bandwidth, the more noise come through the system; hence, the noise power increases.

We could run through the receiver-noise-power equation with any combination of system noise figures and bandwidth values but for our purposes, a narrower range of values will highlight most system combinations. In terms of noise figure, we can examine two possible values: 3 dB and 1 dB. The 3-dB figure represents a realistic value for a suitably designed RF amplifier system using low-noise bipolar transistors, while 1 dB represents what can realistically be obtained using GaAsFET devices in the RF amplifier change. For bandwidth, we'll look at two values: 30 kHz and 250 kHz. The 30-kHz figure represents a nominal IF bandwidth for a basic polar-orbiting-satellite receiver; 250 kHz is included because it represents a reasonable value for the wideband mode in the newer scanning receivers, and is also the minimum value required for reception of signals from the Japanese GMS satellite. If we use this range of values, we can calculate two values for each combination—the receiver noise power level (TNL) and the signal-to-noise ratio (S/N), obtained by comparing the TNL value with the available signal power level (– 134.4 dBm):

Noise Figure	*System Bandwidth*	*TNL*	*S/N*
1	30 kHz	–128.2 dBm	–6.2 dB
1	250 kHz	–119.0 dBm	–15.4 dB
3	30 kHz	–126.2 dBm	–8.2 dB
3	250 kHz	–117.0 dBm	–17.4 dB

As we might expect, a lower value for noise figure and/or bandwidth, does improve (lower) the receiver TNL value. The problem is, the signal is so weak, that

even with a 1-dB noise figure and 30-kHz bandwidth, the signal ends up 6.2 dB *below* the noise-power threshold. With a 3-dB noise figure and a 250-kHz bandwidth, the signal is 17.4 dB below the noise threshold!

In order to display an image with minimal noise effects, we require a signal-to-noise ratio of at least +10 dB. Since our best-case scenario begins with a S/N of –6.2 dB, we obviously need at least a 16-dB boost in signal level because we've run out of options for reducing the receiver noise level. You might think that what we need is simple amplification, but additional amplifier stages amplify the signal power *and* the noise, and won't improve the relationship of the two. The needed boost in signal level—without "benefit" of additional noise—is available in the form of antenna gain!

Impact of Antenna Gain

Antenna options have been discussed in Chapter 2. Our calculation of ground-signal levels assumed no contribution from antenna gain. In practice, of course, you'll be using a gain antenna (and gain comes easily at 1691 MHz!), so the gain of your antenna can be *added* to the satellite-signal power level, thus improving the signal-to-noise ratio. Figure 3.7 plots the effect of antenna gain on system S/N for various combinations of receiver system noise figure and bandwidth. The nomogram covers antenna gain from 0 (with the same S/N values we calculated above) to 35 dBi—about equivalent to that of a 15-foot parabolic dish antenna.

A few levels on the S/N scale are worth noting. A level of 0 dB represents the point where the noise and desired signal have equal power levels. This point is somewhat academic, for although you could clearly hear the signal at 0 dB S/N, the signal would have far too much noise to be usable. The first really critical point on the S/N scale is the +10-dB signal level. This is the threshold level that would provide a noise-free display at the nominal maximum WEFAX power output (+54.4 dBm) from the transmitter. Under conditions of full-WEFAX power output, any combination of receiver noise figure, bandwidth, and antenna gain that is equal to, or above, this level will provide noise-free pictures. This signal level is an appropriate one for use with the GOES Central satellite (WEFAX only) under normal operating conditions. It is also a suitable level for use with the Japanese GMS satellite or the European METEOSAT satellite.

The major complication is the current operational practice with WEFAX from GOES. When the GOES program was first formulated, high-resolution transmission and WEFAX transmission were always scheduled in a non-overlapping manner, resulting in full-power output when the satellite was in the WEFAX mode. The satellites are now much busier, with additional systems, research operations, and time-intensive high-resolution scanning during severe-storm intervals. The result is that now, as a rule, the eastern and western satellites (engaged in active imaging) are now transmitting other data simultaneously with the WEFAX signal. The net result is that WEFAX signal levels are typically 6 dB *lower* than is the case with full-power WEFAX. Noise-free reception under such conditions thus requires a S/N of +16 dB, indicated by the WEFAX + VISSR threshold in Figure 3.7. If we want a system designed for worst-case GOES WEFAX reception, it would be best to use this threshold as your design target.

Putting It All Together

In addition to basic antenna gain values, Figure 3.7 also includes several specific antenna configurations, including parabolic dishes of 0.6 m (2 ft.), 1.2 m (4 ft), 1.8 m (6 ft), 2.4 m (8 ft), 3.0 m (10 ft), and 3.6 m (12 ft). Also included is a 45-element loop Yagi, an antenna that is popular because of its compact size and availability from several vendors.

Assuming a 30-kHz bandwidth for our basic satellite receiver, full-power WEFAX broadcasts could be received essentially noise-free with any of these antennas, although the 0.6-m dish and loop Yagi would yield very little *gain margin* (gain in addition to that required for noise-free display). Some gain margin is almost essential (see below), so of the two smaller antennas, the loop Yagi would be superior. A 1.2-m dish yields a solid gain margin (9.5 to 11.5 dB, depending upon noise figure) for full-power WEFAX reception and represents the practical minimum for reduced-power WEFAX operations. With a 3-dB noise figure, you'll have a slight bit of noise on the image, so the 1-dB system would be the conservative choice with this antenna. In effect, a conservative GOES or METEOSAT installation can be constructed around a 1.2- to 1.8-m dish (4 to 6 feet) with a 1-dB noise figure preferred for the smaller antenna.

Note that if we use a 250-kHz bandwidth, none of the antennas with less than 25 dB gain will yield noise-free display, even at full power WEFAX. To achieve an essentially noise-free display with such a receiver, a 3- to 3.6-m dish (10 to 12 ft) is required. This is the primary reason why I won't be promoting wideband receivers for GOES or METEOSAT reception.

If your station is located in the western Pacific, Asia, or the Australian, New Zealand area, your WEFAX objective will be the Japanese GMS satellite. This satellite transmits using a much wider FM deviation than either GOES or METEOSAT, and a receiver with 250-kHz bandwidth is absolutely required. The only saving grace in this situation is that WEFAX from GMS is almost always at full power, allowing you to design your station around the +10-dB signal threshold. Such an installation requires the use of at least a 1.8-m (6-ft) dish, and a system with a 1-dB noise-figure would be best at the low end of the antenna-size range.

Complications

All of this discussion to this point has assumed no other sources of signal loss or noise in the system. But there are transmission lines and connectors between the antenna feed point and the downconverter RF input. Losses in cables *add* to the system noise figure, so use of a minimum length of quality, low-loss cable is mandatory. Using improper connectors, or improperly installing good connectors, can create losses and further degrade system performance by creating input impedance mismatches that increase the noise figure of the first RF amplifier stage. Water or moisture in transmission lines or connectors radically increases losses, so the precautions noted in Chapter 2 are mandatory to avoid a gradual degradation of performance following initial installation. One of the major limitations of the loop-Yagi antenna designs is that the feed point on this antenna is very critical, yet it is exposed and susceptible to moisture pickup and corrosion—both of which quickly destroy the effectiveness of the antenna.

All of the previous calculations assume optimum matching of polarization at both ends of the circuit. In practice, this is most closely approximated with optimization of a linear feed horn as described in Chapter 2. If a circularly polarized feed is employed, system performance for any combination of factors will be degraded by 3 dB, but polarization won't have to be optimized. If polarization is 90 degrees out of phase (cross-polarized) with a linear feed, losses may exceed 20 dB. If this occurs, you would probably not even hear the satellite signal with any of the equipment combinations we have discussed.

Noticeable reductions in WEFAX signal levels are possible if the ground control station has not acquired the satellite properly with their own antennas, or the uplink power is reduced for any reason (WEFAX signal power is a linear function of the uplink power level with the GOES transponder). Small variations in signal level as a result of such factors are to be expected.

Also keep in mind that at low look angles (cases where the satellite is close to your horizon), result in longer path lengths (and hence greater pass loss) as well as greater potential for partial attenuation of the signal by foliage or other obstructions. Although normal weather variations have negligible effects, a very heavy rainfall can attenuate the signal to some degree.

The bottom line is that almost any additional factors can degrade the signal to some extent, while virtually nothing can happen to increase the signal level! Given

Figure 3.6—The Microwave Modules 1691-MHz downconverter. This particular model is manufactured in the United Kingdom and can be ordered direct or through dealers in various parts of the world. Spectrum International (Concord, MA) markets these units in the US and each unit is individually checked by Spectrum's owner, John Beanland, before each unit is shipped. Spectrum also maintains complete repair facilities, saving North American customers the trial of an overseas shipment should work ever be necessary. The unit features an N connector input for the transmission line from the S-band antenna and a BNC output connector for the 137.50-MHz output signal. A switch is provided to select 1691 or 1694.5 MHz because METEOSAT uses both of these frequencies. Although packaged in a die-cast aluminum case, a weatherproof enclosure is required for remote mounting at the antenna. The smaller unit resting on top of the downconverter is an accessory preamplifier providing a noise-figure of less than 1.5 dB at 1691 MHz.

this oppressive situation, it makes sense to design some gain margin into your system.

When evaluating converter performance, do not simply fix on the noise figure of the first RF stage in the system. I have constantly talked about *system* noise figure. The first RF stage can set the system noise figure only when one or more low-noise stages are used, providing sufficient gain to have the major influence in setting the system noise figure. The system noise figure will always be greater than the noise figure of the first RF stage. The difference can be minor in the case of a multistage, low-noise amplifier, but will increase if only a single low-noise, low-gain stage is employed. When in doubt, try to get the vendor to quote a system noise figure rather than the noise figure of the first RF stage.

Construction

Home construction of a downconverter is a project best suited to folks with a background in UHF and microwave construction. It is not that the projects are complex, for they can actually be deceptively simple. The real problem is that you typically have to have access to fairly sophisticated equipment in order to align and adjust your new creation. If you know what you are doing, extremely good results can be obtained at low cost, but you must be prepared to tinker.

Emiliani and Rhighini (*QST*, November 1980) describe a METEOSAT converter that has the virtue of using readily available, low-cost transistors, but the devices have relatively poor noise figures, and a number of stages of amplification are required because the gain of each stage is quite low. This particular converter was used with a 6-ft (2-m) dish to get the required gain margin. Those seriously interested in construction of a converter should follow technical articles in recent Amateur Radio publications such as *QST* and *The ARRL Handbook*, looking for descriptions of converters designed for use on the 1296- and 2304-MHz amateur bands because such designs can often be converted to 1691-MHz service. Those articles often delve into the design of the converters, something that is critically important if you are going to make modifications to published circuits.

Commercial Converters

Unlike the situation only a few years ago, a number of companies have affordable, off-the-shelf converters available for WEFAX service and new ones, including improved versions of existing models, will appear. Examples of commercial equipment with which I am familiar include the SDC-1691B from Quorum Communications, Spectrum International (Figure 3.6), Microwave Modules in the UK, and Wrasse Elektronik in West Germany. These suppliers also have suitable preamplifiers available, including some using microwave GaAsFETs with noise figures below 1.5 dB! The SDC-1691B from Quorum Communications is an example of a fine state-of-the-art downconverter that combines a number of very desirable features at a reasonable price ($449). The unit features integral GaAsFET preamplification with a 1-dB noise figure. The LO chain is maintained within 1 °C down to a temperature of –20 °C, assuring that the output signal from the converter won't drift significantly over a wide ambient temperature range. This unit is also available with a weatherproof enclosure, thus simplifying installation.

Any downconverter is a significant investment, with prices typically in the $300 to $1000 range. Write suppliers for current model and pricing information

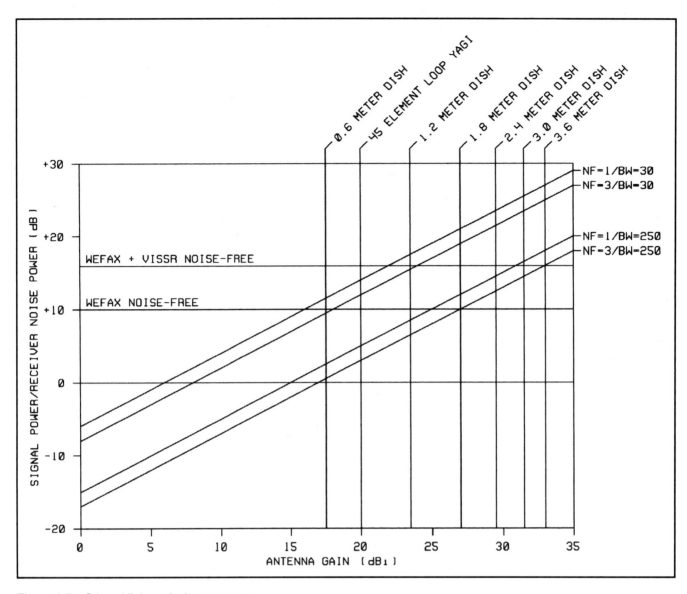

Figure 3.7—S-band link analysis nomogram.

as things change quickly in the dynamic world of precision electronics.

Weatherproofing

I noted earlier that the simplest approach was to mount the downconverter at the antenna. This requires some attention to weatherproofing, even with converters housed in supposedly weatherproof enclosures. If the converter is already weatherproof, it's still useful to provide a ventilated housing designed to keep the converter from direct exposure to the elements. Converters that are not protected by special housings require that you provide an external housing. Considering your investment in a downconverter, this job should be approached very carefully.

Industrial-grade housings, complete with weatherproof seals, are available and can be used in conjunc-

tion with weatherproof type N fittings to make a suitable enclosure. Other enclosures can be fabricated, but be extremely careful that your system is really watertight, for if water gets into a sealed enclosure, it can do more damage than if you used a ventilated housing designed to avoid direct water contact. Cables should be routed with drip loops to avoid water running down cables and soaking connectors. The use of waterproof sealing compounds, designed for TVRO satellite installations may be helpful as well. Unless you can be absolutely certain of the integrity of your enclosure, periodic inspections are a must.

Frequency Drift

The output frequency of a converter is determined by its LO stages and the frequency output can be expected to change with temperature. This change will be small with a well-designed converter, but in the case of some converters, may exceed many tens of kilohertz with seasonal temperature changes at the antenna. The thermal stability of a converter is one of the things that you'll want to check when comparing equipment specifications from various vendors. A converter can be protected from high temperatures by shading the enclosure. Insulating the enclosure also limits thermal excursions.

If your converter drifts with temperature extremes, the receiver will require retuning unless the IF bandwidth is quite wide. If power can be routed to the antenna site, you may want to consider heating an insulated enclosure to some constant value above the highest normal ambient temperature. This will assure a constant output frequency, irrespective of the outside temperature. Such an installation may require the services of a qualified electrician to assure that you meet local code requirements, not to mention the common sense dictates of safety.

A subtle effect of temperature changes in some converters involves changes in the LO-injection level. Some converters may begin to deliver a noisy signal at extremes of temperature. The LO chain in some converters may actually quit at temperature extremes. If your signal is consistently noisy at high or low temperatures, or disappears entirely, the need for thermal stabilization may be indicated.

Another alternative, if the antenna is not located too far from the building housing the receiver, is to use a low-noise preamp at the antenna and house the converter inside, away from the effects of weather and temperature extremes. Since most converters feature low-noise amplifiers at the 137.50-MHz IF, almost any length of transmission line can be used to connect the converter to the receiver elsewhere in the building.

Chapter 4

Video Formats and Display Systems

WEATHER-SATELLITE VIDEO FORMATS

Although working with various satellite video signals is certainly the fastest way to become acquainted with the details of the pictures transmitted by each type of satellite, the learning process is accelerated if you have some prior knowledge about what you should be seeing. If your system display works well from the beginning, you'll have few problems. If there are difficulties with your video circuits, knowing in advance what you *should* see can help you unravel your display problems.

Subcarrier Modulation

As noted earlier, the signal you'll hear from your receiver is a 2400-Hz audio tone. Although the tone maintains the same frequency at all times, it appears to be warbling, or unsteady. That's because the amplitude (volume) of the tone changes according to the different video values on the incoming signal. The tone has a definite maximum peak value that corresponds to the white portions of the image. Although the tone amplitude never quite gets to zero, it can drop to about 4% of the maximum amplitude—the level representing black objects in the image. Values between the black and white limits (4% and 100% amplitude, respectively) represent various intermediate tonal values between black and white.

Displaying the image properly requires that we simultaneously perform two tasks. First, we must somehow convert the varying subcarrier (tone) amplitudes to their equivalent brightness values: a process known as *demodulation*. The goal is to change the display brightness in step with the amplitude changes in the subcarrier. In the case of a TV-like display, the brightness of the spot that is scanning the face of the tube must be altered. In a facsimile system, we must somehow vary the relative brightness of an image that is being printed on paper. All of these devices ultimately are controlled by a range of voltage that must somehow track the subcarrier amplitude changes.

The video circuits described in Chapter 5 provide one example of how this can be accomplished. A typical video circuit often involves some early *audio*

filter stages, designed to pass the 2400-Hz subcarrier signal while rejecting noise and other anomalous signals at other frequencies. Such filtering can never be perfect, but can noticeably improve the quality of the final image when the signal from the receiver is less than perfect.

Filter stages are typically followed by one or more *detector* stages, then additional filtering to remove all traces of the original subcarrier signal, leaving only a dc voltage that varies in value in step with the changing subcarrier amplitude. Because any display circuit requires a certain range of voltage change to go from black to white, a video circuit also requires some gain-control stages to assure that the output voltage has the right value to drive the final display circuits.

In the case of the scan converter that is described in the next two chapters, we want a peak video voltage of +5 for white, dropping to essentially 0 V for black. All the format diagrams in Figure 4.1 assume that we are looking at the output of a video circuit where the maximum output level represents white and the minimum level is black. The actual voltage values vary with the type of display used, but for our purposes, we can assume a 0 to +5-V output level.

All the satellite signal formats covered in this volume have a similar modulation format. Where they do differ from one another is the timing involved in sending the video data. Our display system must precisely time the display process in step with the incoming image. If the timing is slightly off, the image may be tilted or skewed. If timing errors are greater, you'll see no organized image, simply a hopeless melange of light and dark.

In the sections that follow, you'll see how the video information is organized for each satellite type. In each case, the diagrams in Figure 4.1 are analogous to what you might see if you looked at the output of a perfect satellite video circuit using a quality oscilloscope that was precisely in step with the incoming image data. To simplify the diagrams, that part of the signal that represents the actual image is assumed to be an 8-step gray scale, running from black at the left margin to white on the right. When reading the discussions that follow, refer occasionally to the images

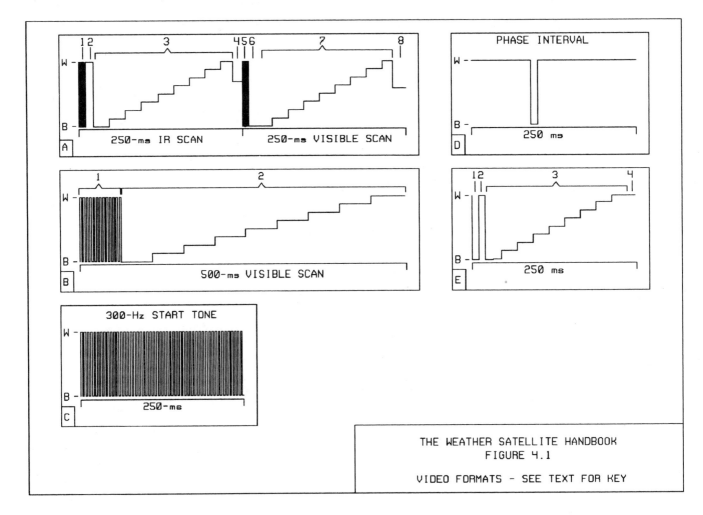

Figure 4.1—Video formats.

illustrated throughout this *Handbook*; they'll help you understand how each satellite video signal is organized and how that relates to the image produced by your display.

TIROS/NOAA APT Format (Figure 4.1A)

The multispectral radiometer of the operational NOAA satellite rotates at two revolutions per second, or 120 revolutions per minute. This rate of rotation is very precisely controlled. One of the major tasks of any display system is to see that the timing on the display stays precisely in step with the spinning radiometer on the distant satellite. One line of composite image data is transmitted with each revolution of the radiometer, so the basic line rate for these satellites is two lines per second, or 120 lines per minute (LPM). Because there are two lines each second, and each second represents 1000 milliseconds (ms), the basic line of NOAA data is 500 ms long.

During the first half of each revolution, while the radiometer is actually scanning the earth, the on-board computers insert the infrared (IR) data. During the second half of each revolution, when the scanning optics are pointed away from the earth, the same computer inserts the visible-light information into the video data stream. Thus, each 500 ms line of image data actually is made up of two different types of image data—IR information for the first 250 ms, and visible-light data for the second 250 ms. Though the majority of each 250-ms interval is taken up with earth scan data, there are a number of other distinctive elements to each line that can help you know what kind of information you are seeing at any moment.

IR Sync Pulse (1)

Each IR line begins with seven cycles during which the subcarrier level swings from white to black and back to white. These seven transitions occur at a rate of 832 Hz. It's possible to design equipment that will detect this sequence of seven 832-Hz pulses, triggering the display of an IR line, but we won't use this pulse in any of our display system in this *Handbook*. This train of pulses appears as a fine series of vertical black-and-white bars if the sync pulse is visible on the display. It is the IR line-sync pulse (and a similar visible-light pulse that will be described shortly) that provides the very distinctive *tick-tock* sound to the NOAA APT signal when it has reached full quieting in the station receiver.

IR Pre-earth Scan (2)

Just before the sensor begins its scan of the earth, there's a brief moment where it is viewing empty space. Because cold is represented by white in the IR data format, this view of space appears white and looks like a white stripe down the edge of an IR image. Once each minute, the satellite clock inserts minute markers into this pre-earth scan zone of the image. These minute markers appear as thin horizontal black markers going almost completely across the white, pre-earth scan. These black markers can provide a valuable time reference during the read-out of the image, while the white pre-earth scan itself is an infallible indicator that the video data to follow represents IR information.

IR Earth Scan (3)

Most of the 250-ms IR line interval is taken up with the actual IR scan of the earth's surface. Warmer objects appear darker than colder objects. If the display is set up for optimum rendition of visible-light data, typical IR data will be strongly biased toward the white end of the output voltage range, producing a low-contrast image made up mostly of whites and light grays.

IR Telemetry (4)

At the very end of the IR line is a zone devoted to telemetry information concerning various aspects of the satellite system. The information is coded as various step-like changes in brightness, resulting in a strip down the right edge of the IR image made up of little gray-scale increments.

Visible-Light Sync Pulse (5)

The visible-light segment immediately follows the IR line sequence and, like the IR line, it begins with a seven-pulse sync sequence. These pulses occur at a 1040-Hz rate, so equipment designed to detect the pulse sequence can respond to either IR or visible sync data. Because the visible-light pulse rate is 1040 Hz, compared to 832 Hz for IR, the visible-light pulses creates a sequence of vertical black-and-white stripes down the left edge of the image that appear slightly narrower than their IR equivalents.

Visible-Light Pre-earth Scan (6)

Because this pre-earth scan represents visible light, the narrow view of space appears black (in contrast to the white of the IR format). Minute markers are also inserted into the black, visible-light, pre-earth scan

interval, but these markers are white (as opposed to the black markers used in the IR format).

Visible-Light Earth Scan (7)

Most of the 250-ms visible-light interval is taken up by the actual earth-scan data. Because this information is essentially equivalent to what the eye would see, it is typically much easier to interpret than the corresponding IR data, particularly if you are new to the arcane business of interpreting satellite data. Clouds will obviously be varying shades of white, water typically appears almost black, and ground features are varying intermediate gray shades. The precise range of gray-scale values depends on the intensity and angle of the sunlight on the earth below the satellite.

Visible-Light Telemetry (8)

The visible-light line ends with a telemetry window similar (but not identical) to the corresponding one at the end of the IR line.

The foregoing description has assumed that we were dealing with a daylight pass during which both IR and visible-light data are available. At night, you'll encounter one of two possible scenarios. The visible-light line interval may be completely dark, or the satellite ground-control stations may direct the on-board computer to insert data from another IR channel, in which case *both* line segments will contain IR data. Because the IR sensors are sensitive to different regions of the IR spectrum, the two views are typically different in appearance.

If the display system operates at 120 LPM, the IR and visible-light data are displayed side by side. This is rarely a good option because the great difference in IR and visible-light contrast ranges makes it difficult to get good rendition of both line segments without elaborate video processing. Unless your application demands a comparison between IR and visible data, it is usually best to run the display at 240 LPM, displaying either the visible-light or IR data and "blanking out" the other data segment. This allows the entire display image area to be devoted to either the visible-light or IR data, greatly improving the resolution of the displayed image.

There are a few subtleties about the APT format that might eventually occur to you. The first concerns image orientation. Any image, displayed as received, has the sync pulse and pre-earth scan to the left of the image data. If you're sharp-eyed, you'll note that in some of the images in this volume, the sync and pre-earth scan are off to the right! These images represent *ascending* or south-to-north passes (see Chapter 8) in which the original image had south at the top (start) and north at the bottom (end) of a display sequence. Most of us prefer to see familiar geographic features presented with north at the top, so many display systems have the facility to invert such images to make the display "look right." When the image is inverted, the sync and pre-earth intervals are on the right instead of the left!

The second subtlety involves *geometric distortion*. Imagine for a moment that you are seated in a glass airplane. You can look out to the left at the horizon, rotate your head until you are looking straight down, and then continue the movement until you are looking out to the horizon on the right. When looking straight down, features are easily recognized, but the view to either side is distorted by the foreshortening of the horizon. Because the scanning radiometer is viewing the earth in a similar manner, the image *should* be distorted toward both horizons. Try as you will, you'll not see such distortion in the images shown here. You would have seen it with the old ITOS/NOAA satellite series (NOAA-2 through NOAA-5), but it is absent from current TIROS/NOAA images because of the activity of that busy little on-board computer. By skipping the actual horizon and varying the rate at which the remaining image data is sent to the APT transmitter, the effects of geometric distortion are largely removed, leaving an image that looks like a simple photograph.

120-LPM Meteor Format (Figure 4.1B)

Compared to the complexities of the NOAA satellite-image format, the Soviet Meteor signal seems simple. Standard Meteor images are transmitted at a rate of 2 lines per second (120 LPM) so that each line is 500 ms long. The start of the line is marked by 13 very prominent white-to-black transitions that make up the Meteor line sync pulse. On a properly displayed Meteor image, this pulse train appears as a series of vertical black-and-white bars along the left edge of the picture. Although the bars usually appear black and white, occasionally it seems as if you can see the underlying cloud cover through the bar pattern. In any case, the sync-pulse train is very prominent and easy to recognize. The remaining line interval is taken up with the visible-light image data.

Until recently, the typical Meteor satellite was strictly a visible-light system, and internal switching would turn the transmitter off when light levels dropped below a critical threshold. It is useless to listen for one of these satellites at night because the transmitters are always off. Descending passes on winter mornings are sometimes a bit surprising. The satellite can be above my horizon to the north, but the satellite's transmitter is still off because of the low light levels from the polar landscape below. As the satellite moves south, light levels rise; at some point, the transmitter suddenly turns on. In such a case, the signal seems to appear out of nowhere at full-quieting, in contrast to the usual gradual reduction in noise as the satellite rises above the RF horizon.

For the last year or so, the Soviets have been experimenting with what appear to be IR transmissions. Most of these early tests had scanning rates of about 20 LPM, and little success was achieved in decoding the format. As I write this, Meteor 3-3 is testing a new 120-LPM IR format. These images have the typical Meteor sync pulse, but look a little odd when displayed. This is because the actual video values are *inverted*: The clouds and cold objects are dark, while warmer objects are white. To make such an image conform to the more familiar IR format (cold = white and warm = black), the video must be inverted or complemented. In an analog system this requires an extra video-inversion stage (usually an op amp); in digital systems, the complementation can be performed by mathematically manipulating the video data (see Chapter 7).

240-LPM COSMOS Imagery

As noted in Chapter 1, Soviet COSMOS satellites are occasionally heard using a 240-LPM video format with standard subcarrier modulation. These satellites are in lower orbits than typical Meteor or TIROS/NOAA satellites, so the passes tend to be shorter and the area covered by an image is considerably less than you are used to seeing. This smaller coverage area, together with the 240-LPM scanning rates, makes for spectacularly detailed pictures. Aside from the scanning rates, these transmissions can be identified by columns of numbers along one edge of the image. Such satellites are heard only occasionally and they may tend to have short operational lifetimes. In any case, they are a real treat if you happen to catch one.

Occasionally you'll also hear 240-LPM COSMOS transmissions that result in very unusual displays. These seem to be test vehicles for Soviet radar satellites.

WEFAX Signal Formats (Figure 4.1C-E)

TIROS/NOAA and Meteor transmissions are continuous, resulting in a strip of imagery built up throughout an entire pass. Any given picture from such a satellite simply represents a segment of the original transmission. In contrast, WEFAX transmissions (from the GOES) are individual pictures or frames that typically require 200 seconds for display. The transmission format for a specific frame is rigidly organized in ways that make it easy to automate the display of WEFAX transmissions. The individual images are transmitted at a rate of 4 lines per second (240 LPM) using the standard subcarrier modulation format. Thus, each WEFAX line has a duration of 250 ms.

Start Tone (Figure 4.1C)

A WEFAX image (or frame) begins with square-wave modulation of the subcarrier (between the black and white limits) at a frequency of 300 Hz. This produces a very distinctive tone from the receiver and, if viewed on a display system, produces a total of 75 paired black-and-white vertical bars across the width of the display. This start tone interval is 5 seconds long in the case of the US GOES satellites, and 3 seconds in duration for the European METEOSAT and Japanese GMS satellites. The start tone can be used to automatically activate a display system at the start of a frame transmission.

Phasing Interval (Figure 4.1D)

The start tone is followed by an interval in which the subcarrier is at the white level, interrupted by short (about 12-ms) pulses where the subcarrier drops to the black level. These black pulses occur four times each second, and mark the point where the image lines begin in the upcoming picture. These pulses act as markers that allow the display system to get in step with the image—a process known as *phasing*. Some additional notes on phasing are included at the end of this chapter. The GOES phasing interval is 20 seconds long, while the METEOSAT and GMS satellites have 5-second phasing intervals. Because the black phasing pulse is easily differentiated from the adjacent white-level data, it is relatively easy to implement an automatic phasing system.

Image Transmission (Figure 4.1E)

Once the phasing interval is finished, the actual transmission of image data begins. There is a narrow, vertical, black bar at the extreme left edge of the line (1), a narrow white-level framing interval immediately to the right (2), the image video data (3), and a narrow white border (4) to the extreme right end of the line. The nominal WEFAX frame consists of 800 lines. Since the lines are transmitted at a rate of 4 lines per second (240 LPM), the required transmission time is 200 seconds. Transmissions of Tropical East and West quadrant images are slightly shorter and that of some weather charts is a bit longer.

Stop Tone. When the frame is complete, the subcarrier is modulated between the black and white limits at a frequency of 450 Hz. Like the start tone, the stop tone is easily recognized and is used, in automatic systems, to disable the display system at the end of the picture. The duration of the stop tone is about 5 seconds.

DISPLAY SYSTEMS

There are three basic approaches to the display of weather-satellite pictures: Use a slow-scan CRT monitor, a facsimile recorders or various forms of scan conversion. Each of these will be reviewed briefly.

CRT Display

Direct, cathode-ray tube (CRT) display of the weather-satellite formats represents one of the most-

easily visualized and easy-to-implement forms of image display. The operation of a CRT system has much in common with conventional TV display. In a conventional TV set, an electron beam is scanned across the face of the tube at the proper horizontal rate, creating a "scanning line." As the beam moves across the screen, the intensity of the beam is varied, thus varying the intensity of the glow created by the phosphor on the face of the CRT picture tube.

At the same time the beam is scanning horizontally, it is also moving vertically, so each line that is written on the face of the tube falls just below the previous line. The combination of vertical and horizontal scanning, in combination with the varying brightness of the trace, paints a complete picture on the face of the tube. The type P4 phosphor on the face of a standard black-and-white TV tube has a very short persistence, which means that it does not glow very long after the electron beam passes over a given point on the screen. The factor that gives the picture the illusion of permanence on the screen is the very fast rate at which the picture is painted. The face of the tube is scanned horizontally over 15,000 times each second, and a complete image is formed every $\frac{1}{60}$th of a second. The eye simply cannot perceive such rapid changes, and the brain integrates all the information and interprets the face of the tube as containing a complete picture.

In principle, a CRT monitor works the same way. In the case of a WEFAX picture (in which lines arrive at the rate of 4 each second over the 200 seconds required to create an 800-line picture) if we scan the tube horizontally at the 4-Hz rate, scan it vertically from top to bottom in 200 seconds, and cause the trace to lighten and darken in response to the variations in subcarrier amplitude, we can paint the WEFAX picture on the face of the tube. Unfortunately, the picture information is being painted on the tube so slowly (compared to a standard TV picture) that we cannot see all the image at one time. Using a P4 tube, all we can see is the current line being scanned. If we were to use a tube with a long-persistence phosphor, such as those used for radar displays (phosphor types such as the P7), we might be able to retain up to a few dozen or more lines on the screen—if we view it in a dimly lit room—but we would never see the entire picture at once. We can, however, take a time exposure photograph of the screen over the 200-second frame period, and the film, when processed, will contain the picture.

Monitors that work directly at the satellite scanning rates thus must involve photography to enable us to see the picture in its entirety. This is a major limitation of the CRT approach, but there are some very positive points that have made CRT displays quite popular. First, because the various satellites differ primarily in areas such as scanning rates (and scanning rates can be altered with simple circuit modifications), a CRT monitor

Figure 4.2—A CRT display system designed and constructed by Grant Zehr, WA9TFB. (*Reproduced from* The 1990 ARRL Handbook, *Fig 30, p 28-18*).

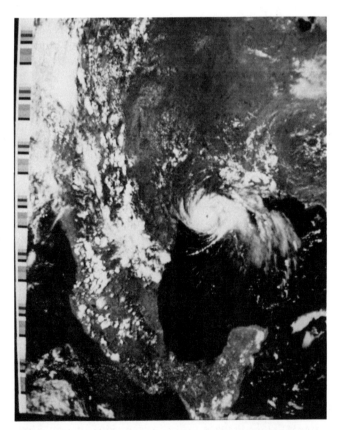

Figure 4.3—Display of a NOAA-7 visible-light image (17 August 1983) showing hurricane Alicia making landfall on the US Gulf coast. (*Reproduced from* The 1990 ARRL Handbook, *Fig 37, p 28-22*).

is easily constructed or altered to handle virtually any image format. Constructing such a monitor is fairly simple. Making it work is not much of a chore because you can directly observe all of its functions or malfunctions and easily comprehend what is occurring.

It is this simplicity and flexibility that have made CRT displays quite popular among weather-satellite enthusiasts *despite* the fact that you have to photograph the screen. Your major chore, should you choose to take this approach, is in acquiring specialized parts such as the CRT tube, high-voltage power-supply components, and deflection yokes. Considerable latitude is possible: Creative salvaging of parts from small-screen black-and-white TV sets, or the use of off-the-shelf replacement parts for such models, will often suffice.

An excellent example of a CRT display system, designed by Grant Zehr, WA9TFB, can be found in *The 1990 ARRL Handbook* (pp 28-13 to 28-22), and is illustrated in Figure 4.2.

Facsimile (Fax) Recorders

A fax recorder is a useful combination of electronic circuits and mechanical features that creates a direct print of the satellite image. The print can be created using light on sensitive photographic paper, which produces a permanent image when processed, or it can be created immediately by various kinds of electrolytic or electrostatic recording papers. The major advantage of fax is the fact that you have a direct and

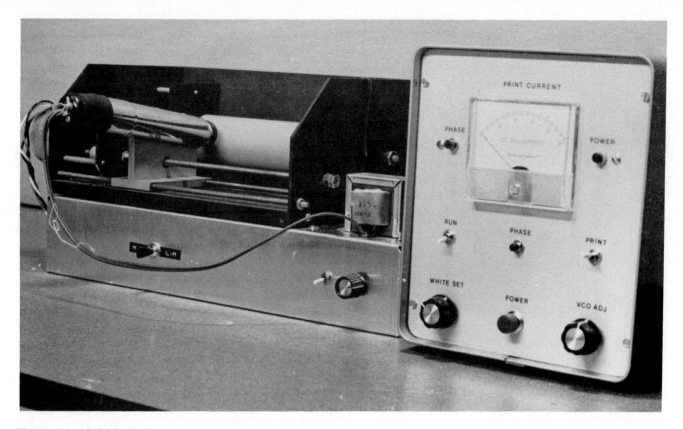

Figure 4.4—A home-built drum facsimile recorder constructed by the author. The control and drive electronics are housed in a metal cabinet and the various signals are routed to and from the drum and motor assembly with a multiconductor cable. The mechanical assembly includes precision synchronous motors for drum drive (240 r/min) and for driving a threaded rod that moves the printer carriage (40 r/min). This particular system is set up for direct printing on sheets of photographic enlarging paper so printing is via a light gun made up of an R1168 glow modulator tube and a lens assembly from the eyepiece of a microscope. The lens is required to ensure that the modulated light source is focused as a fine spot of light on the paper wrapped around the drum. The electronics assembly contains video circuits similar to those in Chapter 5, a high-voltage driver circuit (300 V maximum) for the glow modulator tube, a crystal-controlled 60-Hz frequency reference for drum speed control, a 60-Hz power amplifier and step-up transformer to provide the 120 V ac required by the drum motor, and a variety of circuits for control and phasing.

immediate (or almost immediate) printout of the picture. The prints are typically of fairly large size and can be extremely detailed.

Unlike a photographic print, which can be made small or very large, the fax recorder image is limited to a single size because of the mechanical nature of the printing system. This mechanical factor also effects flexibility with regard to the different modes of satellite operation. The scanning rates in a fax recorder are determined by motor speeds, gears, and mechanical linkages. So, changing scanning rates is fairly complicated compared to the modest circuit changes needed to achieve the same flexibility with the electronic scanning circuits of a CRT system.

Whereas a CRT system is almost all electronics, a fax recorder involves a considerable amount of mechanical machinery in the form of motors, drums, drive mechanisms, etc. A home-built fax recorder is a tinkerer's delight, and can range from a real trial (for those who hate all things mechanical), to a delight for those who like to work with precision machinery. With everything working just right, a fax recorder can produce the best possible images with the least bother, but a well-built CRT system comes very close to this—with considerably less effort.

The drum of the fax recorder must operate at a very precise speed. This requires the use of a synchronous motor driven by an audio power amplifier driven by a crystal-controlled reference. Video circuits can be similar to those shown in Chapter 5, but which output circuit is employed depends on the type of recording paper used.

Photographic paper or film requires the use of a modulated light source used in conjunction with a lens

Figure 4.5—A photographic fax printout of a GOES WEFAX frame. This image was printed using the fax recorder illustrated in Figure 4.4. The light source was a crater lamp with a lens system constructed from microscope optics. The system was set up to print directly onto photographic enlarging paper, producing a 7- × 7-inch image.

system to focus a sharp point of light on the surface of the photographic material. High-intensity light-emitting diodes (LEDs) might be suitable if sensitive photographic film is used as the recording medium. In such a case, the video circuits must provide maximum brightness on white, and minimum brightness on black. If photographic papers are used (eliminating the need to process the film and then expose and process the final print) a brighter light source (such as an R1168 gas-crater tube), or a solid-state laser diode, must be used. In either case, the white subcarrier must generate minimum light output, while black results in maximum lamp output. Because a photographic recorder uses light-sensitive paper or film, the recorder must be operated in the dark (in the case of most films), or under suitable safe-light illumination (papers) until processing is complete.

Fax machines using electrostatic papers typically apply a modulated voltage (45 V for white, up to about 300 V for black) to a wire stylus. Electrolytic papers are commonly used with blade and helix continuous-read-out recorders. In such designs, printing voltages in the 75- to 150-V range are employed.

The major problems in the construction and operation of a fax recorder is the steadily rising cost of new synchronous motors, and obtaining a supply of the specialized papers for other than photographic systems. Figure 4.4 shows one of my drum-type fax recorders with a sample of photographic output presented in Figure 4.5.

Digital Scan Converters

Digital scan converters offer an option that was not even practical when the first editions of this *Handbook* went to press. They are practical today because of the proliferation of microcomputers and the drop in price and increase in capacity of memory chips. Basically, a digital scan converter samples the incoming picture and converts the video values to numerical values that are stored in digital memory. Although the picture samples are slowly added to memory as reception proceeds, they can be read *from* memory very quickly. In fact, it's possible to read a set of picture samples so quickly that you can create a steady picture on the screen of a standard TV set or video monitor.

Being able to view weather-satellite pictures on a TV monitor has great appeal. If we add the ability to reorient the picture, zoom in on details, or process the contrast range of the picture—all possible once the image data is in a computer's memory—scan converters look like real winners. In fact, they are, and I am watching two different monitor displays as I write this.

Because the scan converter samples and stores video information, limitations on available video memory control how detailed an image we can produce. A practical scan converter does not yet have the inherent full-resolution capability of something like a fax recorder, but the pictures from the most basic scan converter can match the resolution capabilities of a standard TV monitor, allowing you to watch thousands of excellent satellite images without using up film or fax paper. Modern computer graphics displays can provide essentially full-resolution display and the cost of such capabilities is dropping rapidly.

DIGITAL–VIDEO BASICS

In general terms, digital scan conversion involves the use of digital techniques to process a picture transmitted with one set of scanning standards so it may be displayed on a system with different standards. The most common—and most difficult—application of digital scan conversion to weather satellites is the processing of pictures so that they may be displayed on a standard TV set or video monitor. At first glance, this seems to be an almost impossible task given the inherent differences between fax and conventional TV. Fax formats are designed to transmit very-high-resolution still images. To keep the bandwidth narrow, the picture must be transmitted rather slowly: four lines per second for WEFAX, for example, requiring 200 seconds for the transmission and reception of an 800-line image.

By contrast, conventional TV deals with images of moderate resolution that must have the capability to reproduce motion. In order to do so, individual pictures must be transmitted at a speed high enough to cause the brain to merge the train of incoming images into an impression of smooth motion. In the simplest form of conventional TV, known as *non-interlaced* scanning, a 262-line image is transmitted in $\frac{1}{60}$th of a second, with individual lines arriving at the rate of 15,750 each second!

A second, somewhat more complicated format, used for commercial TV broadcasting, is *interlaced* scanning. This format involves the transmission of a 262.5-line scan in which the first scan, requiring $\frac{1}{60}$th of a second, paints the even lines of the image, followed by another scan, again requiring $\frac{1}{60}$th of a second, for the odd-numbered image lines. The odd lines are inserted between the even lines (hence the term *interlaced*), yielding the complete image, built up from the two interlaced scans, every $\frac{1}{30}$th of a second.

Interlacing has the advantage of increased resolution (512 lines compared to the 262 lines of the non-interlaced format) while minimizing flicker in the display. Although there is a definite doubling of resolution in the interlaced format, few TV sets or monitors are kept in precise enough adjustment to realize the benefits. Our discussion of scan conversion will concentrate on non-interlaced operation of the TV monitor, both for simplicity of signal generation and conservation of memory. The latter is a subject we'll examine shortly.

Principles of Scan Conversion

Our ability to perform digital-image scan conversion is based on the remarkable properties of computer memory chips. These chips store numerical values and make them available at a later time. If we sample a satellite image as it arrives (oh, so slowly!), converting the varying brightness values of the picture to their numerical equivalents, these brightness values can be loaded in sequence into solid-state memory. Once the image values have been stored, they'll remain available unless we purposely overwrite them with new values, or turn the system off. Although we loaded the image values slowly as they arrived, we can read them out very quickly—and do so as often as we want. If we time the readout properly, we can convert the image numerical values into a conventional TV picture. Because we read the entire picture out of memory, again and again, every 1/60th of a second, what you see is an image that remains on the TV screen until we load in new image data or turn off the power.

Image Sampling and Memory

Up to this point, we have always dealt with satellite video in an analog context where we could look at the video signal as a continuous stream of subcarrier data in which the subcarrier amplitude could be essentially *any* value between the 0% and 100% amplitude limits. Digital techniques involve the storage of *specific* numerical values for specific points in the image so we must *sample* the picture by breaking each line into specific units known as picture elements or *pixels*. The average brightness level of the subcarrier during the time interval represented by each pixel must be converted to a specific numerical value—a process known as *analog-to-digital* (A/D) conversion. How finely we can specify the brightness of an individual pixel is a function of how many *bits* of memory we want to devote to the storage of individual pixel data. Each bit of memory can store information in binary (0 or 1). The number of gray-scale values that can be coded with n bits of memory can be calculated by raising 2 to the nth power (2^n):

Storage Bits	Gray-scale Steps
1	1
2	4
3	8
4	16
5	32
6	64
7	128
8	256

Computer memories are organized as 8-bit units called *bytes*. If we use one byte of memory (8 bits) for every pixel, each pixel can be specified to one of 256 possible gray-scale values. In terms of tonal resolution, this is very-high-quality video, and the human eye is unable to differentiate this many tonal steps in a black and white display. Indeed, most people have difficulty distinguishing 64 gray-scale values on a TV display (6 bits/pixel). In terms of satellite display, many fax papers have difficulty in adequately differentiating even 16 gray-scale steps, something you can prove to yourself by printing out the 16-step gray scale that is sent out once each day on the GOES-Central schedule. Sixteen gray-scale steps per pixel produces a reasonably nice TV display, but anything less has a very definite tonal contouring that reminds most people of a paint-by-numbers picture.

Sampling Resolution

How good the picture will look is really a function of two factors: the precision of the gray-scale coding for each pixel, and how many pixels or samples are collected to reproduce the picture. If we use the WEFAX format as an example, the picture is inherently limited in vertical resolution by the 800 scanning lines. Horizontal resolution is a bit more complex. Recall that the signal is transmitted as AM on a 2400-Hz subcarrier. If we use full-wave video detection, the resulting signal is carried on a 4800-Hz tone. If our post-detection filtering were absolutely perfect, the maximum useful resolution could never exceed 4800 samples per second. That's because of the maximum resolution inherent in the original subcarrier signal. Since there are four WEFAX lines per second, the theoretical maximum resolution of a 240-LPM signal is 1200 pixels per line. Because video filters are not perfect, sampling the signal to yield 1024 pixels per line is convenient in digital terms and yields a practical resolution that exceeds that available on most analog CRT and fax systems.

If we sample 1024 pixels on each line for a total of 768 lines (a more convenient value in the digital world than the 800 lines that are actually transmitted), we require a storage capacity for 1024 × 768 pixels—a grand total of 786,432 pixels per frame! If we devote one byte for each pixel (8 bits per 256 gray-scale steps), we would need a digital memory with a capacity of 768 kbytes! This exceeds the 640 kbytes of memory that can be directly addressed with standard 16-bit machines such as the IBM PC and its many clones. And, it's over 10 times the total memory capacity of many of the 8 bit microcomputers such as the Color Computer or the standard Apple II!

If we are willing to reduce tonal resolution to 16 gray-scale steps (a quite acceptable level when coupled with high spatial resolution) we can store two pixels per byte (4 bits per pixel), reducing the display memory to something in the order of 384 kbytes. You can

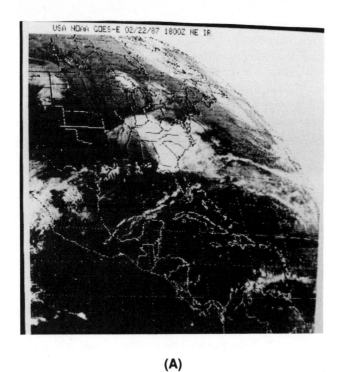

(A)

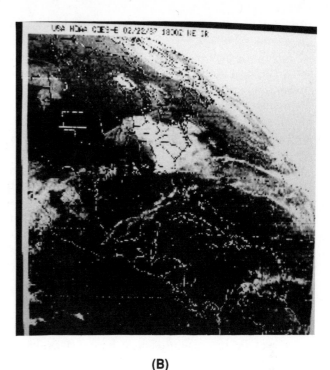

(B)

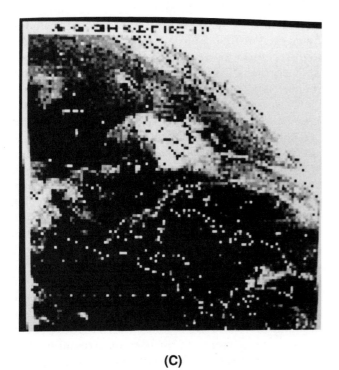

(C)

Figure 4.6—The effects of spatial resolution on a GOES-E NE IR image quadrant. All images were displayed on a 640 × 480 VGA computer display. Sixteen gray-scale step coding (4-bit) is used for all images. (A, upper left) shows the image displayed at 512- × 480-resolution. Although this is only half the resolution that the scan converter can supply in the streaming data format (see Chapter 7), the image is quite clear. All elements of the ID header are clearly rendered, as are geographic coordinates and political boundaries inserted by the NOAA ground-station computers. (B, above) shows the same image, but at 256 × 256 resolution, equivalent to what the scan converter provides in the stand-alone mode using a composite TV monitor. Note that the header data are noticeably degraded, although the header is still readable if you are familiar with the format. Many details of the coordinates and geographic outlines have been lost although enough remains to be useful. In (C, left), the resolution has been dropped to 128 × 128. Although you would still recognize the limb of the earth and be able to say that the image was an IR quadrant, all useful printed and coordinate information is lost.

obtain off-the-shelf VGA cards and monitors for IBM PC-compatible computer systems that display such 1024 × 768 × 4-bit imagery directly, and comparable resolutions can be obtained with Amiga and Macintosh computer systems. Although still somewhat costly, degrees of resolution are now practical. The IBM "standard" for the VGA display is 640 × 480 × 4 bits,

and most of the satellite images in this book were obtained using such a system (see Figure 4.6A).

If picture display on a standard TV monitor is your goal, display memory requirements can be much less exacting and still provide useful results. If we examine the non-interlaced TV display, we are limited to something in the order of 262 lines. In terms of efficient use of memory, it's often convenient to look at the display of a 256-line picture. If the vertical resolution of the TV display is limited to 256 lines, there is not much to be gained by sampling more than 256 pixels on each image line. The total number of pixels in the TV format would thus be 256×256, or 65,536 picture elements. If we use 1 byte for each pixel, we need 64 kbytes for storage of a non-interlaced TV image. If we limit the gray-scale steps to 16, allowing us to store 2 pixels per byte, our memory requirements can be reduced to 32 kbytes, a requirement we can meet with a single memory chip! Although lacking in precise details, such a display is still useful and has the advantage of low cost (see Figure 4.6B). Any less spatial and tonal resolution results in an image with comparatively little value, as shown by the 128×128 image in Figure 4.6C.

System Configuration

There is a comparatively wide range of functions that must be implemented in a basic image-display scan converter:

- Memory management, including proper addressing of memory in read and write operations, timing of control pulses, etc.
- A/D conversion to transform satellite video data into numerical values for storage.
- D/A conversion to convert the digital data in memory back to analog data for display.
- Precision timing for TV picture display.
- Detection of line sync or clock signals to control pixel sampling.
- Provisions for proper phasing of the incoming image.
- Control of line sampling.

All of these functions require a certain amount of hardware, but there is a variety of ways in which the control functions can be implemented.

Hard-wired Control

It is possible to perform all of the scan-converter control functions using analog circuits, timers, and a variety of logic configurations. The various logic and control functions are basically wired into the system, depending upon the input and output functions desired. The major disadvantage of this approach is that if you want to add or change formats, you must change the control function hardware. Depending upon your initial approach, such modifications can range from simple to impossible. In all cases, since control functions are hardware dependent, the circuitry of hardwired systems can be quite complex and is always more complicated than some other alternatives.

Computer Control

Many aspects of the operation of a scan converter represent a sequence of actions that you could perform with a series of simple switches—if you could react quickly enough. As an example, if you had a set of instructions for loading WEFAX data and suitably fast reflexes, you could function as follows:

1) Wait for a line sync pulse.
2) Get a video sample:
   A. Store sample to memory
   B. Update memory address
   C. Wait n timing cycles
   D. Was this last sample #256 for the line?
   E. If yes, set sample count to 0 and go to #3
   D. If no, update count by 1 and go to #2
3) Wait for a total of 2 line pulses
4) Was the last line number 256?
   A. If yes, stop loading
   B. If no, update line count by 1 and go to step 1

This set of instructions represents a combination of functions: operations to be performed—getting samples, storing values in memory, updating counts, and decision-making steps. The decision making includes things such as "if a count has reached a certain value, do this—otherwise do that. . . ." If we can break down these decisions into simple expressions, we have the essence of a computer program. Computers are certainly fast enough, particularly with assembly-language programs, and it is a *relatively* simple matter to interface the computer to the outside world in a control context. In effect, with a proper interface, virtually all scan-converter control functions can be implemented by a computer program. One of the major pluses of this approach is that the system operates as a function of its programmed instructions. If we want to add new capabilities or change existing ones, we can accomplish this largely by altering the instructions. Often, no hardware changes are required. In all cases, any such changes will be trivial when compared to a hardwired system.

Any computer-controlled scan converter represents two major system components: the electronic hardware and the operating software. Although there is a tendency to concentrate on hardware requirements (which can range from the simple to the complex) the real stumbling block is almost always going to be the software. Although easy-to-use, high-level languages such as BASIC suffice for menus and disk file load and save operations, the actual image-handling routines must operate at a far higher speed than BASIC can manage. Generally, this means that critical routines must be written in assembly language, or in a fast version of compiled languages, such as C. Unless you are already proficient in one of these languages, you may base your scan-converter decision on the availability of the necessary software.

Perhaps the easiest way to implement a scan converter is to use a microcomputer such as a PC clone, Amiga, or Mac. If the computer has high-resolution graphics capability, the additional hardware required can be as simple as some video and clock circuits and interface hardware for the computer bus. In the case of computers with relatively low-resolution graphics, external hardware to provide a gray-scale image at moderate resolution would be desirable. In recent years, many companies have come out with multimode computer interface units, some of which have HF or satellite fax capability. Many of these units work quite well with specific computer systems when used to display weather maps, but a bit of caution is in order in evaluating their performance with the gray-scale images we are interested in.

A typical weather map can be displayed using binary output (a pixel is either black or white), so a relatively good-looking map can be displayed on most computer systems and the maps can be printed on most dot-matrix printers with good results. When it comes to gray-scale images, you may discover that the hardware or software supports only a limited number of gray-scale steps, or your particular computer may not be capable of providing a reasonable gray-scale image. Gray-scale images printed on typical dot-matrix printers are marginal at best. They may be interesting if you have never seen a satellite image before, but the excitement fades quickly if you compare the printed image with that of a picture from a system optimized for gray-scale image display or printout.

There are a great many potential systems available, and I have not had the opportunity to examine more than a handful of them. A number of hardware and software vendors and their products are listed in Appendix I. In evaluating a system for possible use, you should look for the highest possible combination of spatial and tonal resolution. Sixteen gray-scale steps is certainly the minimum number for useful work. In terms of spatial resolution, you should be able to display a minimum of 200 lines, with at least 250 pixels per line. Obviously, more is better, as you can see from Figure 4.6.

There are two general problems with microcomputer-based scan converters. First, the computers generate varying amounts of RF noise that may present problems with VHF reception. Some computer models are worse than others, and you may want to check with people using such computers if you are planning on purchasing a specific computer for your system.

The second problem involves possible conflicts of use, especially with a relatively expensive computer. Obviously there is no problem if you can afford to dedicate a computer to satellite display but, for most

Figure 4.7—The WSH scan converter. All of the electronics are housed in a compact 4 × 11 × 7-inch cabinet. Operated in a stand-alone mode, the scan converter displays images from any of the operational weather satellites on a standard TV monitor. The scan converter is also easily interfaced to a computer to save and load images from disk. If there is sufficient resolution in the computer display system, the scan converter provides an option for higher-resolution image display, image processing, and a host of other advanced functions. Shown here with a 5-inch TV monitor and the Vanguard WEPIX 2000 receiver, it is obvious that a basic ground station can be extremely compact. Because all of these items operate from a 12-V dc supply, this entire installation could be operated from a remote site with little difficulty.

of us, an expensive computer must be dedicated to more than one task.

Another major approach to computer control involves the use of a scan converter with a dedicated microprocessor. Such a unit has, in effect, its own built-in computer, and operates from software programmed into a read only memory (ROM) chip. Such units can be quite compact, generating a TV image with mode selection and other functions controlled by front-panel switches. Such units have two major advantages. First, they generate little or no bothersome RF noise because the computer circuits are in the same shielded cabinet with the other scan converter hardware and there are no interconnecting cables carrying digital data. Such cables are a major source of RF noise in a microcomputer-based system. Secondly, they require no accessory computer and thus are ideal if you have no computer. The primary disadvantage of such units is that you are usually limited to the functions that have been built and programmed into the unit—a

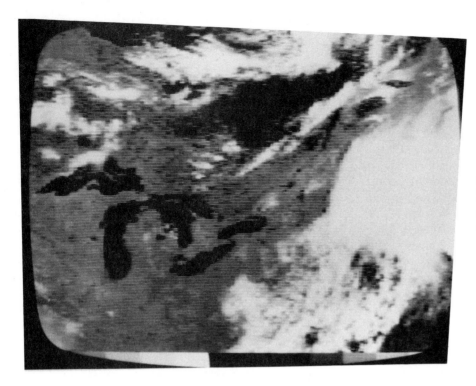

Figure 4.8—An example of the 256 × 256 resolution provided by the scan converter in the stand-alone mode in conjunction with a conventional TV monitor.

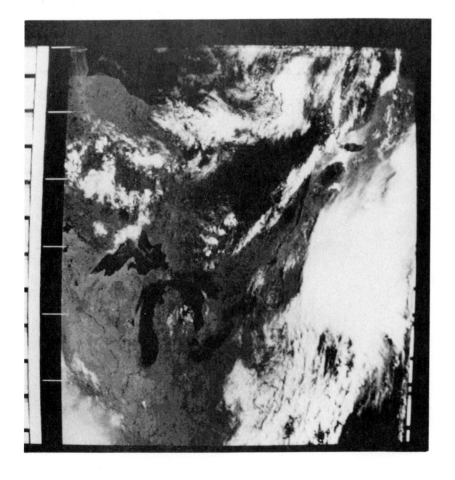

Figure 4.9—Using a computer with high-resolution graphics capability in conjunction with the scan converter provides the option for higher resolution display, as shown here with a 512 × 480 VGA display of the same image shown in Figure 4.8. Though this version of the image is obviously more detailed than the one provided directly from the scan converter, all of the major features of this image are resolved in the stand-alone version. The major disadvantage of the image in Figure 4.8 is a loss in image coverage caused by a combination of overscan on the composite monitor, and loss of data during the vertical- and horizontal-blanking intervals.

fact that may or may not be limiting, depending on the specific circuit.

The scan converter (Figure 4.7) to be described in the next two chapters combines the best of both the dedicated and microcomputer options. The unit is a complete stand-alone scan converter that generates a 256 × 256, 16-gray-scale step image using a conventional TV monitor, or a TV set in conjunction with an RF modulator. It is easy and inexpensive to build, and doesn't require elaborate test equipment to put it into operation. It displays all satellite video formats in real time (as they're being received), or from audio-tape recordings. In effect, the unit is an ideal entry-level system since it is a basic black box that only requires a TV monitor to round out your basic station display system.

In addition to its stand-alone capabilities, the unit also incorporates two parallel ports, enabling it to be easily interfaced to virtually any microcomputer. Simple communications parameters are programmed into the ROM, greatly simplifying data exchange between the scan converter and an external computer. Image transfer can be accomplished in BASIC, if desired, and even the simplest computer system can thus be used to save and load images from the scan converter using disks or tape.

For those of you who have computers with advanced graphics capabilities, the scan converter "streams" high-resolution data in a 1024 × 768 × 6-bit (64 gray-scale step) format, giving you the advantages of high-resolution image display and a wide range of options for image processing. Even if your computer lacks high-resolution graphics, you can still capture the high-resolution data stream, process the images as desired, and pass the processed image back to the scan converter for display!

The built-in communications protocols also permit you to control all scan-converter functions from the computer keyboard, if desired. In contrast to a microcomputer-based system, however, if you need to use the computer for some other purpose, the scan converter will still function in the stand-alone mode, so you needn't miss pictures while you are working on another project. This is particularly convenient when receiving WEFAX images because up to several hundred pictures may arrive each day, and the scan converter displays incoming WEFAX images automatically.

Chapter 5 describes the built-in computer for the scan converter as well as the satellite video circuits. Chapter 6 covers the display and timing electronics and testing and packaging. Chapter 7 is devoted to interfacing the scan converter to an external computer and how to go about programming your system to take advantage of the many functions an external computer can provide, including some basics on image processing. The operation of the scan converter is covered in Chapter 9.

Total costs for the scan converter need not exceed $250. Printed-circuit boards and a pre-programmed ROM chip are available to make the project easy to duplicate. Add an external computer to the system, and you open the options for almost unlimited expansion without any modifications to the scan converter.

Chapter 5

The WSH Microcontroller

INTRODUCTION

In the previous chapter, I introduced the concept of digital scan conversion. In order to properly format an image for TV display, the satellite video signal has to be sampled and passed to the display memory. All of this must be precisely timed. There also have to be provisions for: detecting the start of WEFAX images, determining proper signal phasing, responses to operator input to stop or start the conversion process, the display of images from memory, inversion of the displayed images, and a host of other possible functions. There are three basic approaches to obtaining these control functions:

- The use of hard-wired logic circuits
- The use of a microcomputer
- Integration of a microcontroller

With the hard-wired approach, all of the needed control functions are implemented with specific circuits—timers, counters, detectors, logic arrays, etc. The range of possible control functions is determined by how all of these various elements are connected, with specific interconnections required for each of the possible modes of operation. The hard-wired approach can be complex to implement—particularly with a wide range of modes and functions. It also has the disadvantage that adding new modes or functions can be difficult (or even impractical), depending upon the initial design. For all but the simplest functions, the hard-wired approach is definitely limited.

The use of a microcomputer, such as a PC, offers considerable flexibility. Some interface hardware is required to allow the computer to sense and control various functions associated with the scan converter, but the range of possible functions is largely a matter of the operating software. If the computer has sufficiently high graphics resolution, the computer itself can function as the scan converter with very little additional hardware.

As powerful as this approach is, there are some things you must take into account. First, there's the choice of computer. Given the specificity of interface hardware and operating software, a given system can, at best, be implemented within a single family of computers—such as the IBM PC and its many clones. Users of other types of computers may have to extensively redesign given interface hardware and write new software in order to implement a comparable system.

Then, there are the concerns of cost and possible conflicts for use of the computer (who's going to use it for what job at what time?). Although the various scan-conversion functions must be performed very quickly and with great precision, these tasks are relatively trivial for powerful, modern microcomputer systems. A computer with the graphics capability to provide a quality image display can easily cost more than $1000. Even if you can afford such a computer, you may not want to dedicate it exclusively to image display.

A third consideration involves RF-noise generation that can interfere with the reception of the VHF satellite signals. The level of RF interference created by some computer systems may be completely unacceptable; even a well-shielded system can sometimes significantly desense your receiver, usually interfering with reception early and late during a satellite's pass.

All of these concerns and questions can be avoided by using a controller to operate the scan converter. A *controller* is nothing more than a simple computer on a small PC card. Typically, they use a *microcontroller* chip (often in the Intel 8xxx or Motorola 6xxx families) as well as a range of support chips including memory and input/output (I/O) functions. Controllers are widely used in industry for monitoring and process functions. Even today's appliances and automobiles have become dependent on the versatility that such controllers provide.

There are dozens of available controllers on the market. Most begin to approach the $300 to $400 price range when configured for the job of controlling a scan converter. Choosing a specific controller to do any job takes careful consideration. For instance, what if the device goes out of production after you're committed to undertake the project? To avoid this prob-

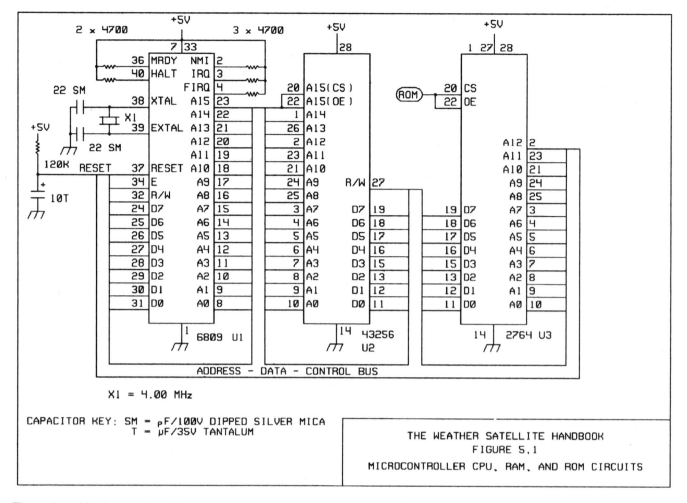

Figure 5.1—Microcontroller CPU, RAM, and ROM circuits.

lem, I designed a controller specifically for this project: it's built around the Motorola 6809 *microprocessor.* This particular processor is, in my own view, one of the most powerful and versatile of the 8-bit units, and provides a cost-effective approach in making the scan-converter controller decision. Because the controller is small and—along with the other scan-converter electronics—is packaged in a shielded cabinet, RF noise is essentially eliminated.

Lest you feel intimidated by the construction of even a simple-minded computer, rest assured that the task can be painless! You don't need to know anything about computers to get the system up and running. If you want to use the circuit as a learning tool, it's ideal—it lends itself to a host of other applications, if you take the trouble to learn how it functions. Using the available PC board (see Chapter 6) you can build the controller in just an hour or so with a total component investment of under $50.

When combined with the display circuits described in Chapter 6, you'll have a completely functional scan

converter that displays all current satellite-image modes on a standard TV monitor, or on a TV set (with the use of a simple RF modulator that's readily available from video stores or other outlets). If you have a computer system, it can be interfaced to the self-contained scan converter with little additional hardware or software (see Chapter 7). This permits you to save images on disk, perform image processing, and use a very-high-resolution display (if your computer has that capability). Virtually any computer can be interfaced to the scan converter.

Use of a stand-alone scan converter is ideal because you don't need to have (or use) a computer in order to watch satellite images on your TV display. If you already have a computer (or later acquire one), you can steadily upgrade the capabilities of your display system without having to make major new investments in anything but computer hardware. In effect, the scan converter provides a simple entry into satellite-image display. It has the capability for almost unlimited expansion by simply upgrading an external computer you may have. Because it can stand on its own, the scan converter is ideal for demonstrations, portable operations, or emergency use because it operates from a single 12-V dc power source.

Using the available circuit boards and software, the cost of the complete scan converter ranges from $250 to $300, depending upon your parts source, choice of cabinet, etc. For those who do not want to take the time to build the system, wired and tested scan converters are available.

The remainder of this chapter is divided into three sections: a circuit description of the controller, construction, and a basic programming guide. The latter is not a necessity because software is available, but it's included to assist those who want to experiment with the system, or use the controller in other applications. A description of the display circuits that complete the scan converter project are included in Chapter 6. Chapter 7 is devoted to the hardware and software aspects of interfacing the scan converter with a wide range of microcomputers. Chapter 9 includes a complete guide to operating the scan converter, based on the available software.

CIRCUIT DESCRIPTION

The controller circuits are most easily understood in terms of the following functional elements:

- The 6809 microprocessor
- Address decoding
- System ROM for nonvolatile program storage
- *random access memory* (RAM) for image storage and storage of variables during program execution
- A bi-directional 8-bit port with control strobes for communication with the scan converter display board
- Parallel 8-bit input and output ports for communication with an external computer
- Satellite video circuits and analog-to-digital (A/D) converter
- Power-supply components

Each of these functional elements will be discussed in the sections that follow.

Microprocessor (Figure 5.1)

The 6809 microprocessor (U1) is a typical 8-bit microprocessor; it outputs or receives data via 8-bit directional data lines (D∅-D7). These 8 data lines comprise the system *data bus*. The processor also has a total of 16 address lines (A∅-A15). There are 65,536 possible logic combinations of this address bus (16^2), so the processor is said to have a 64-kbyte address space. The 64-kbyte address space accessed by the *address bus* is allocated among several different functions, a subject

to be discussed under the *Address Decoding* discussion that follows.

All timing functions for U1 are derived from an internal clock oscillator whose frequency is controlled by a 4-MHz crystal (X1). The actual operating clock frequency is 1 MHz and, although this may appear slow compared to the 33-MHz PCs now in service, given the flexibility of the 6809, it is more than fast enough for any of the functions that must be performed.

The 6809 has a variety of interrupt and control lines (IRQ, FIRQ, NMI, MRDY, and HALT) that are not used in the scan converter. All of these lines are disabled by pulling them high with 4700-ohm pull-up resistors. One interrupt that is used is the system RESET line. This line is connected to an RC network comprised of a 120-kilohm resistor and a 10-µF tantalum capacitor. It takes the RESET line slightly more than one second to go high on power-up before the processor is activated, assuring that all system chips and voltages have stabilized before the 6809 begins to execute its first instructions.

Address Decoding (Figure 5.2)

The 64-kbyte address space of U1 can be used for various kinds of memory as well as I/O ports. Partitioning of this memory space into the system *memory map* is a function of the address decoder (U4). As we shall see shortly, the lower 32-kbyte address space is devoted to RAM. U4 uses address lines A12 to A15 to partition the upper 32-kbyte address space into a total of eight 4-kbyte blocks that are allocated to a variety of functions. In the discussions that follow, addresses are indicated in hex notation because that is the form commonly used for programming.

You need not understand this decoding to build and use the controller, but I've included it for the benefit of those who want to experiment with other applications for the controller, or want a complete understanding of the various scan-converter functions. The complete memory map for the controller is shown below:

F∅∅∅h-FFFFh	System ROM containing the operating software
E∅∅∅h-EFFFh	Spare 2—not used in this application
D∅∅∅h-DFFFh	Spare 1—not used in this application
C∅∅∅h-CFFFh	EXI/O controlling the 8-bit input and output ports for communicating with an external computer (if used)
B∅∅∅h-BFFFh	VREAD—strobe to read video values from the on-card A/D converter
A∅∅∅h-AFFFh	DRESET—reset the display counters

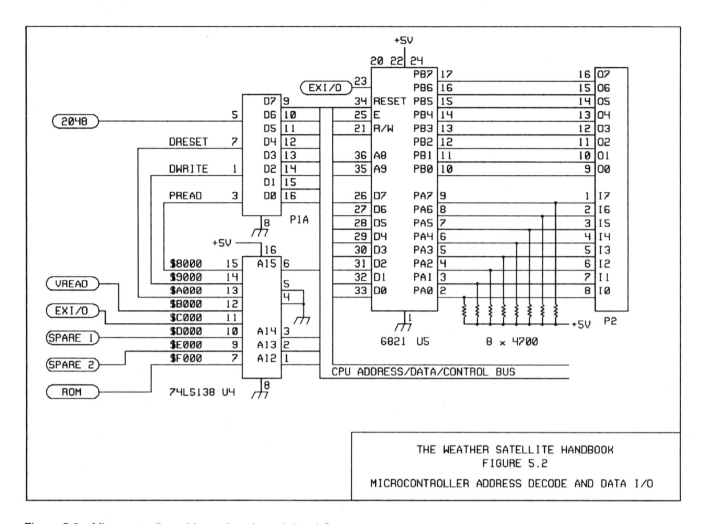

Figure 5.2—Microcontroller address decode and data I/O.

9000h-9FFFh DWRITE—write values to the display memory

8000h-8FFFh PREAD—read switch and clock data from the display board

0000h-7FFFh System RAM

Read-Only Memory (Figure 5.1)

All of the system operating software is stored in ROM using an 8-kbyte × 8-bit EPROM, U3. Although U3 can hold up to 8 kbytes of nonvolatile data, the addressing scheme used here makes only the upper 4 kbytes (F000h through FFFFh) available to the system.

Random-Access Memory (Figure 5.1)

The card has 32 kbytes of RAM in the form of a single 32-kbyte × 8-bit static RAM chip (U2). The Chip Select (CS) and Output Enable (OE) lines are active lows that are connected to address line A15. Only if A15 is low (indicating an address in the lower 32 kbytes of the memory map) will U2 be functional, effectively mapping the system RAM into the lower 32 kbytes of the memory map. The address decoder (U4) functions only if A15 is high, so there is no conflict between the lower and upper 32-kbyte partitions of the 64-kbyte address space. All but 128 bytes of the 32 kbytes of available RAM are used for image storage—enough to handle 255 lines of image data, each containing 256 4-bit pixels. Of the remaining memory, 64 bytes are allocated to the hardware stack, while the remaining 64 bytes are used as storage for temporary data values during program execution.

Bidirectional Data Port (Figure 5.2)

Operation of the scan converter requires that the controller communicate with the display board that is described in the next chapter. The controller must be able to send data to the display to accomplish three tasks: It must reset the display counters when beginning a new image (DRESET), and write data to the display memory (DWRITE). It must also be able to read the status of the various front-panel switches and

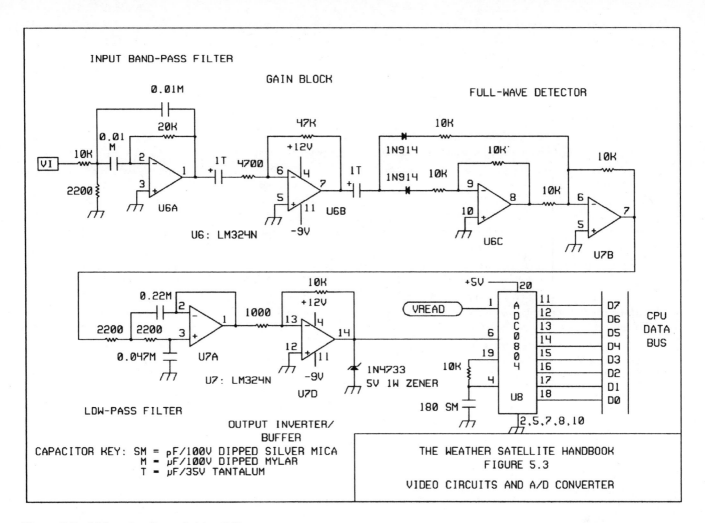

Figure 5.3—Video circuits and video A/D converter

the status of the system clock signal—both tasks are accomplished by the PREAD strobe from the address decoder.

The eight CPU data lines and the DRESET, DWRITE, and PREAD strobes are routed to a 16-connector DIP socket (P1A) that connects to a matching socket on the display board via a 16-conductor ribbon cable. The display board also generates a crystal-controlled 2048-Hz tone that is used for system clocking; this signal is routed to the controller card using P1A. Thus, P1A provides all the necessary connections between the two boards using a pre-wired header cable, greatly simplifying the final wiring of the scan converter.

Eight-bit Input and Output Ports (Figure 5.2)

Although the final scan converter functions fine as a basic TV display without the need of a computer, one of its more powerful features is that it can communicate with an external computer. This ability lets you save images to disk, or take advantage of a variety of

advanced high-resolution display or image-processing functions. The simplest approach would have been to implement a serial port for such communications, but serial data transfer is too slow for passing high-resolution data, even at very high data rates. Instead, a pair of parallel ports—one dedicated to output to the computer and another for input from the computer—are used.

U5, a Peripheral Interface Adapter (PIA) provides the two 8-bit parallel ports. Each bit of the two ports can be independently set for use as input or output, but in this application, the software sets port A as an input while port B is used for output. All eight input data lines to port A are held high by pull-up resistors so the system will function with or without a computer connected to the two data ports. If you ever use this circuit in another application, you have a total of 16 TTL I/O lines available that can be programmed independently for input or output functions. The 16 I/O lines are routed to a 16-pin DIP socket (P2) so that they can be connected to the interface or other circuits by means of a standard 16-conductor ribbon header cable.

The PIA requires access to the eight CPU data lines (D∅-D7), the R/W line, the processor E clock, and the CPU RESET line. The EXI/O strobe from the address decoder (C∅∅∅h) enables the chip and—together with address lines A8 and A9—selects which of the data ports or internal PIA control registers (used in programming the bit status for each port) are selected.

Video Circuits and A/D Converter (Figure 5.3)

In the original version of the scan converter, the video circuits and A/D converter were mounted on the display board. In the latest version of the boards, they have been moved to the controller board to maximize the flexibility of using the controller in other applications. If the video circuits are eliminated, the A/D converter is still available to read analog voltages—a very useful option in many of the possible applications for the microcontroller circuit.

The satellite video signal, taken directly from the receiver or from a tape deck, is applied across a chassis-mounted 10-kilohm CONTRAST control (see the Mainframe Wiring diagram in the following chapter) and is then routed to point VI on the controller circuit board. U6A is an audio active bandpass filter with a bandwidth of 1600 Hz, a center frequency of 2400 Hz, and unity gain. This simple filter stage provides for considerable rejection of noise falling outside of the satellite video passband. U6B is a simple gain block (gain = 5) to provide sufficient drive for a full-wave video detector made up of U6C and U7B. The output of the detector is a negative-going amplitude modulated signal containing a 4800-Hz frequency component as a product of the full-wave detection of the 2400-Hz subcarrier signal. The 4800-Hz component is filtered out, leaving a variable dc signal voltage, using an active low-pass filter comprised of U7A. U7D inverts the negative-going video signal to a positive-going waveform and provides a fixed voltage gain of 10. In use, the CONTRAST control is adjusted so that black signals produce essentially 0 V, while white levels drive the output voltage to +5 V. A 5-V Zener diode is placed across U7D's output to ground to assure that video levels cannot exceed +5 V.

The varying dc voltage (0- to +5-V range) is applied to the input of a hardware A/D converter (U8), which converts the analog input to 8-bit digital output. The VREAD strobe from the decoder (B∅∅∅h) is applied to the Chip Select input (pin 1) of U8. The eight data-output lines from U8 normally float in a high-impedance state that has no effect on the CPU data bus. When the controller wants to read the current video value, VREAD pulls low, and the output value on the eight data lines can be read from the CPU data bus.

The eight digital outputs from U8 represent 256 possible digital values. With a 0- to +5-V input, video values are read with a single-step resolution down to about 20 mV. In other applications, a 0- to 5-V signal at the input to U8 can be read with similar resolution. The A/D stage can be operated in two possible modes—you can trigger an A/D conversion with a read request, but then you must wait approximately 100 microseconds for the result of the conversion. In this application, U8 is wired to make conversions continuously, so that when a read request is made by pulling VREAD low, the results of the last conversion are immediately available, simplifying both hardware and software design.

Power Supply (Figure 5.4)

In order to simplify wiring, the controller board is designed to operate directly from +12 V dc. This voltage provides the positive voltage for the two op amp packages (U6 and U7). The +5 V required by the various digital circuits is obtained from a +5-V regulator chip (U10), which mounts on the controller card.

The op amps also require a negative voltage. Because only a few milliamperes of operating current are required, a simple on-card circuit is used to generate this negative voltage from the +12-V bus. A 555 timer chip (U9) is wired as a free-running oscillator operating at a frequency of approximately 12 kHz. The output of U9 drives a diode and filter network that produces a regulated −9-V output.

Liberal bypassing is provided by 0.1-μF disc capacitors on all voltage buses, with at least one capacitor for every two active devices on the board. A smaller number of 10-μF tantalum capacitors is used to further bypass the high-current +5-V bus. Not all bypass capacitors used are shown on the schematic, but the parts list includes all capacitors required for the circuit board available from Metsat Products (see details at the end of the next chapter).

CONSTRUCTION

A complete parts list for the microcontroller and the display board is presented in Appendix II. This listing covers all parts for the circuit boards, mainframe wiring, and packaging, exclusive of mounting hardware.

Printed-Circuit Board

Construction of the controller is very easy if the Metsat PC board is used. Figure 5.5 shows one of these boards after circuit wiring is complete. The board size is approximately 6 × 8 inches. The PC board is double-sided with plated through-holes and the layout is uncrowded. The documentation provided with the board provides exploded views of the parts layout so proper insertion of parts should be fairly easy. Use sockets for

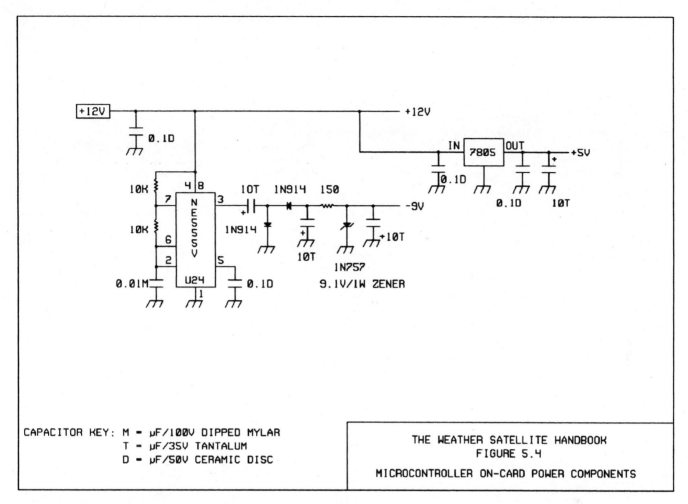

CAPACITOR KEY: M = µF/100V DIPPED MYLAR
             T = µF/35V TANTALUM
             D = µF/50V CERAMIC DISC

THE WEATHER SATELLITE HANDBOOK
FIGURE 5.4
MICROCONTROLLER ON-CARD POWER COMPONENTS

Figure 5.4—Power supply.

all ICs. Defective chips are very rare (from reputable suppliers), but if you get one, it's difficult to remove a soldered-in chip without damaging the board. When you've completed the board assembly, set the board aside for a while. Return to it later for an independent check for proper parts insertion. Be particularly careful regarding the orientation of IC sockets and polarized capacitors and diodes. If you have soldered in a socket incorrectly, *don't attempt to remove it*—simply make sure that you *insert the chip correctly* after you have performed the preliminary tests at the end of the construction section!

Hand Wiring

The circuit can be wired using perfboard and point-to-point or wire-wrap techniques (the prototype was built this way), but the job then becomes far more tedious and error-prone because of the large number of address- and data-bus interconnections. Such hand-wired circuits can develop power-bus noise problems that are difficult to diagnose. Unless you are equipped

and experienced in troubleshooting high-speed digital circuits, use the PC board.

If you do choose to hand-wire the unit, I suggest that you make several copies of the schematics and, using a colored pen or pencil, mark each connection as it is made. Use heavy-gauge wire for all power buses, and use bypass capacitors liberally. When wiring is complete, set it aside for a period. Later, use an ohmmeter and check each socket pin to verify that it is connected properly. This is a tedious job, but it's really the only way to spot a wiring error in a complex circuit of this type. Again, you can use copies of the schematics and colored pencils to verify the connections as you go along.

Initial Tests

There is very little that can be done to test the controller board prior to completion of the display board, but you should check the power circuits prior to inserting most of the chips in their sockets. Insert U9 and apply +12 to 14 V dc (the output of a 13.8-V dc "12-V" supply will do nicely) to the +12-V point with

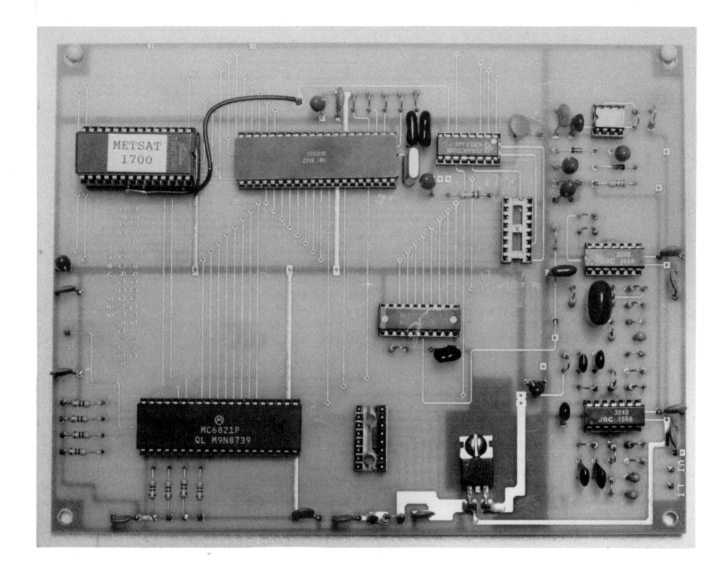

Figure 5.5—A wired controller circuit board. The large IC at the center of the upper edge of the board is the 6809 microprocessor. Immediately to the right of this chip is the microprocessor crystal and associated capacitors, and the 74LS138 address decoder. Just below and to the right of the address-decoder chip is an empty DIP socket (P1) that provides the interface for the display board to be described in Chapter 6. To the left of the 6809 is the 43256 video RAM, with the software EPROM mounted on top of the RAM IC to ease the layout constraints on the PC board. The 20-pin device immediately below and to the right of the center of the board is the ADC0804 A/D converter. The large IC in the lower left is the 6821 PIA chip that provides the two parallel ports for communicating with an external computer. The empty DIP socket immediately to the right of the PIA is P2, providing access to the 16 TTL I/O lines. The card-mounted regulator, which permits the board to be powered from a basic +12-V bus, is located to the right of the DIP socket. The analog video circuits are located along the right end of the circuit board. The two 14-pin ICs are the LM324s used in the video chain; the 8-pin device is the NE555 that provides the negative bias voltage for the LM324s.

the negative output of the supply connected to a convenient ground point. The no. 6-32 screw used to secure the 7805 regulator to the board is an ideal ground point.

The voltage on the +5-V bus should be within the range of +4.5 to +5.5. If it is outside this range (highly unlikely), replace U10. The +12-V bus potential should equal the supply input voltage. On the –9-V bus, you should measure between –8.5 and –9.5V. If it is out of range—or of incorrect polarity—check the placement and orientation of the parts around U9. Pay particular attention to diodes and the 10-μF tantalum capacitors.

Assuming that all voltages check out, power down and insert the remaining ICs, double-checking the orientation of each prior to seating it in the socket. Then, set the controller aside until you have completed and tested the display board described in the next chapter.

SOFTWARE

The controller obviously requires operating software if it is to perform the job of controlling the scan converter. For several reasons, writing software for a project of this sort is a challenge. First, all the code must be written in 6809 assembly language, then compiled (using an assembler program) into the binary instructions and data that the 6809 processor can understand. When the code is written and has reached the point where you get an error-free assembly, it must then be burned into an EPROM using an EPROM programmer and its software. Finally, the code must be debugged as it actually controls the processor. This can be the most challenging phase because the controller does not have elaborate screen displays and printer drivers to tell you what is going wrong if the code does not behave as expected. Successful debugging demands that the programming be second nature and that you be familiar with every aspect of the system hardware. The whole process is an experience that can try the patience of all but the most avid experimenter!

For the vast majority of those who want to duplicate the project, the simplest approach is to buy a programmed ROM chip: you just insert the ROM into the controller. A programmed ROM for this project is available for $50 (shipping pre-paid in the US and Canada) from Metsat Products (see Appendix I). Order the Model 1700 ROM. This chip supports all of the operating features described in Chapters 7 and 9.

EPROM PROGRAMMING
AND SOFTWARE DEVELOPMENT

A complete object-code listing for the ROM is available from the ARRL (see Appendix I). To use the listing to program a ROM chip, you must have an EPROM programmer that handles 2764 EPROMs, and software to input the code and operate the programmer. The hardware and software to accomplish this is available for almost all personal computers, but if you don't already have them on hand, they'll cost you many times the price of the programmed ROM chip.

The software is protected by copyright and it is provided for strictly noncommercial use by individuals. The programming code represents literally hundreds of hours of programming and debugging. If abuses in the form of sale or large-scale distribution of programmed chips emerge, they will be pursued.

You can, of course, develop your own code for the scan converter or any other possible function for the controller; I encourage you to do so if you are experimentally inclined. The programming notes provided in the section that follows are designed to assist you in such a project. You need not pay any attention to this material if you intend to simply buy the pre-programmed ROM, but you may find the discussion interesting if you've had any previous experience in assembly language programming.

If you are seriously interested in experimenting with your own software development, you'll have to become proficient in assembly language programming. If you are already able to program in assembler for another processor, you have two options. The first is to find a commercial controller card that uses a processor identical (or similar) to the one with which you are already familiar. Such a card has to provide (or be modified to provide) the functional equivalent of all the ports provided by the WSH Microcontroller. Then you can develop code for the card and use it with the display board in the next chapter.

The second alternative is simply to get proficient in 6809 assembly coding. This won't take long if you are already proficient in another assembler language because the principles are generally the same and the details won't take long to master. If you have never done any assembly programming, you'll obviously be starting from scratch. A good background in BASIC or another high-level language is a start. Here are some fundamental guides for 6809 programming:

1) *MC6909-MC6809E Microprocessor Programming Manual*, originally published by Motorola Inc. in 1981, with periodic reprints since then. This manual can be ordered through Motorola dealers: Order number M6809PM(AD). This book is the bible for this processor.

2) Zaks, R. and W. Labiak, *Programming the 6809*, Sybex Inc (1982), 2344 Sixth Street, Berkeley, CA 94710, 362 pp.

3) Warren, C. D., *The MC6809 Cookbook*, Tab Books Inc (1980), Blue Ridge Summit, PA 17214, 190 pp.

The third book covers programming considerations for the 6821 PIA as well as the 6809 processor.

You'll also need a programming platform in the form of a computer with software tools for doing 6809 assembly language programming. If you're using an IBM PC or clone, advertisements of companies that market cross-assemblers for the 6809 can be found in *PC Week*, *Byte*, *The Computer Shopper*, and other computer publications. If you're using another family of computers, consult magazines targeting your system to see what might be available.

Perhaps the most inexpensive and appropriate programming platform for the 6809 is the Radio Shack Color Computer because this inexpensive home computer uses a 6809 microprocessor. If you choose this route, W. J. Barden, 1983, *TRS-80 Color Computer—Assembly Language Programming* is an essential reference. It is available at most Radio Shack outlets (RS 62-2077). Radio Shack markets a number of assembler options for the Color Computer. Other assembler programs, as well as EPROM-programming hardware and other options can be found in ads in *The Rainbow* (Falsoft Inc, The Falsoft Building, 9509 US Highway 42, Box 385, Prospect, KY 40059). This monthly magazine caters to the Color Computer and is available worldwide.

A wide range of specialized software and hardware projects for the Motorola 68xxx family of processors can be found in the *68 Micro Journal*, published monthly by Computer Publishing Inc, 5900 Cassandra Smith Road, PO Box 849, Hixon, TN 37343. Although this magazine is available at larger establishments carrying a wide range of computer-related periodicals, it may be difficult to obtain in some areas.

Just for the record, the original code for the scan converter was developed using the Radio Shack Color Computer, the disk version of the EDTASM+ assembler from Radio Shack, and a hardware simulator program that I wrote for the Color Computer. EPROM programming from the Color Computer was handled with hardware and software from Green Mountain Micro. The commercial ROM package has been ported to an IBM PS/2 Model 30 with assembler support from Avocet.

To say that working up such a programming package is a hobby itself is not an exaggeration! Though the time and effort invested are immense, there are few things more satisfying than watching an embedded microprocessor handing a complex task that you have taught it to accomplish.

Chapter 6

Scan-Converter Display Board

INTRODUCTION

Although the microcontroller board provides the intelligence required for the scan converter, the display board provides most of the hardware resources needed to implement the remaining scan-converter functions. If the display board were presented as a single schematic, the circuit would look quite complex. Fortunately, its actual functions are implemented with a series of circuit modules, many of which are quite similar. This makes the circuit fairly easy to understand and build. It has been designed to allow for all initial tests to be done with nothing more complex than a logic probe and a multimeter and all board functions can be verified without the use of the microcontroller. Both the display and microcontroller modules can be completely tested on the bench before installing them in a cabinet.

The following discussions look in turn at the circuit description, construction, and testing of the display board. The final section discusses packaging and final wiring. Complete operating instructions are contained in Chapter 9, Station Operations.

CIRCUIT DESCRIPTION

Like any complex circuit, the display board can best be understood by breaking the design down into individual functional units:

- The microcontroller interface
- Display addressing
- Write-address circuits
- Display data input/output
- Clock circuits

In each case, a paragraph or so of troubleshooting data is supplied after the basic circuit description. This is for your use if later tests of the board reveal a problem in a particular circuit. What is *not* covered in these notes is the obvious first step in troubleshooting—determining that all components are properly

inserted in the board (or interconnected, if you've used point-to-point wiring).

The Microcontroller Interface (Figure 6.1)

The eight data lines from the microcontroller (D∅-D7) and three data strobes (PREAD, DWRITE, and DRESET) are carried on a 16-conductor ribbon header cable that mates with P1B on the display board. All eight data lines are available for writing data to the display memory. The four low-order data lines (D∅-D3) are also used to provide data to the microcontroller regarding front-panel switch settings and clock status. This function is implemented using a 74LS244 bus driver (U29). If there is one IC whose function you need to understand in order to comprehend the various scan-converter circuits, it is the '244.

The 74LS244 is an *octal buffer/line driver*, which simply means that it functions as an 8-pole switch that can either connect or isolate eight data lines from an 8-line data bus. The eight inputs and outputs of the 244 are arranged as follows:

Input Pins	Output Pins
2	18
4	16
6	14
8	12
11	9
13	7
15	5
17	3

The sole function of the '244 is to control whether or not the eight inputs are electrically connected to the eight corresponding outputs. This is determined by the logic status of pins 1 and 19, which function as control inputs. Although the two pins can be actuated independently (each controlling a bank of four inputs and outputs), all of our applications tie the two control pins together so that all control inputs

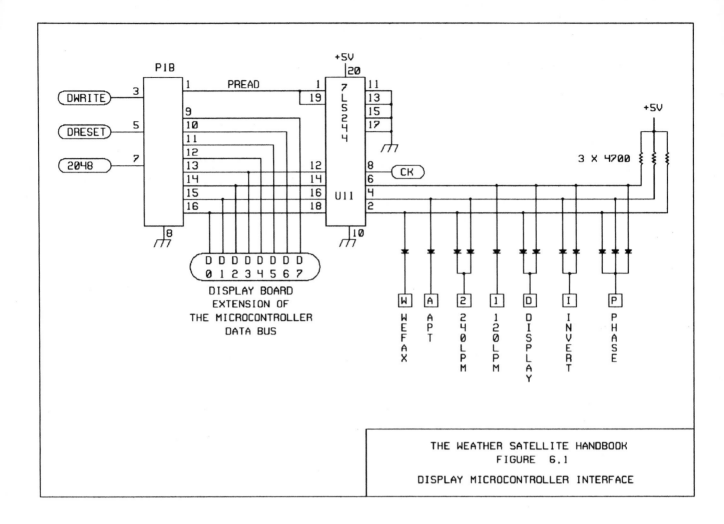

Figure 6.1—The microcontroller interface

affect all eight I/O pairs. If the control pins are high, the eight outputs are held in a high-impedance state in which the input logic levels have no effect. Essentially, if pins 1 and 19 are high, the eight inputs are completely isolated from the outputs, and the outputs have no effect on the data bus to which they are connected.

If pins 1 and 19 are pulled low, the logic state of the eight input pins appears at the eight output pins, and their values are applied to the data bus. In effect, pulling the control pins low transfers the logic values at the eight possible inputs to the data bus. It is this switching function of the '244 that permits a single 8-bit data bus to be used in several different ways. All we need to assure is, that for any given bus, only a single '244 is active at any given moment. If we can accomplish that, the bus can be used to carry a variety of signals at different times.

The switch and clock functions require that only four of the possible eight inputs and outputs of U29

be used. The four unused inputs are grounded to minimize noise transients in the IC. The three low-order inputs are pulled high with 4700-ohm resistors, and the corresponding three outputs are connected to lines DØ-D2 of the controller data bus. The switching function of U29 is controlled by the PREAD strobe. When the controller wants to read the status of the switches, it pulls PREAD low so that the value of the three switch-input lines (pins 2, 4, and 6) appears on lines DØ, D1, and D2 of the data bus.

The three input lines can have 3^2, or 9 possible logic states, all of which are employed in the scan converter. The front panel **MODE** switch has four possible positions: **WEFAX**, **APT**, **240 LPM**, and **120 LPM**. In addition to the **MODE** switch, there are also front-panel, push-button switches for **DISPLAY** (display the image in the controller memory), **INVERT** (display an inverted image from the controller memory) and **PHASE** (phase the incoming image). Diodes are used to connect each switch in a unique way to the three switch data inputs,

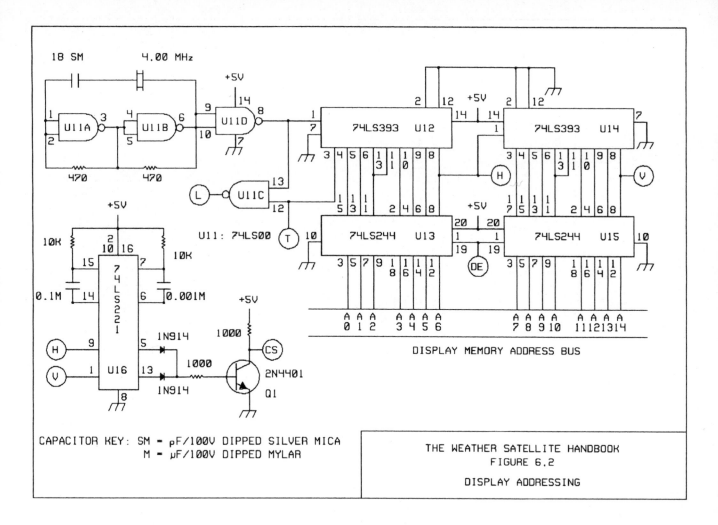

Figure 6.2—Display addressing.

resulting in eight possible values for the three data lines, depending upon which switches are activated:

Switch	D∅	D1	D∅	Decimal Value
HOLD (none)	H	H	H	7
WEFAX	L	H	H	6
APT	H	L	H	5
240 LPM	L	L	H	4
120 LPM	H	H	L	3
DISPLAY	L	H	L	2
INVERT	H	L	L	1
PHASE	L	L	L	∅

Using a combination of how the switches are wired (for example, **DISPLAY** and **INVERT** are only active in the **HOLD** position of the **MODE** switch) and what values are legitimate in certain portions of the program, the controller can read the status of any of the front-panel control switches and respond accordingly.

The scan converter also requires an accurate clock signal in order to time image sampling, execute time delays, and perform other functions. This clock signal (CK) is applied to bit 3 of the U29 inputs, and the controller can read the clock status by executing a PREAD and examining that status of line D3 of the data bus. If this line is high, then the clock is high, and vice versa.

There are many more applications of the 74LS244 in the circuits that follow. The only difference between the function of U29 and these other uses is the fact that we'll typically use all eight input/output pairs, rather than four lines as just described.

If problems are encountered in the I/O area during later testing, remove all the ICs *except* U29. With power applied, use a logic probe to check the status of the eight data-bus lines. With pins 1 and 19 floating (not grounded), you should get no reading on the probe. Use a test lead to ground pins 1 and 19, and check the status of D∅-D3 in sequence as each one of the switch points (**W**, **A**, **2**, **1**, **D**, **I**, and **P**) are grounded, in turn, with a clip lead. If you don't get the logic patterns described above, U29 is defective.

Display Addressing (Figure 6.2)

A 32-kbyte static RAM chip is used for display memory, providing the capacity to store 256 lines of image data with each line composed of 256 4-bit (16 gray-scale steps) pixels. A total of 15 address lines (A∅-A14) is required to access all the memory locations in this chip (2^{15} = 32,768). Most of the time, the chip data is read out at very high speed to generate the TV image. This requires that the address lines cycle serially at high speed, a function accomplished with two dual divide-by-16 binary counter chips (U12 and U14). The counter chips are driven by the output of a 4-MHz oscillator (U11A and U11B), with U11D providing a buffer stage. Each of the counters has eight outputs. The first output of U12 (pin 3) is not used for addressing, but is combined in U11C with the 4-MHz input signal to generate a latch control signal (L) that is used in the data-output section. The 2-MHz signal is also used as a toggle enable (T) that is also used in the data-output stages.

The remaining 15 outputs of U12 and U14 provide the needed 15 lines for display addressing of the memory chip. Although we want to clock the memory at high speed for display, we must have the capability to interrupt this addressing when we want to write new data to the display memory chip. Therefore, we must have a means of disabling the display addressing, something we accomplish by routing the 15 address lines through a pair of 74LS244 bus drivers (U13 and U15), controlled by the display enable (DE) strobe. Normally, DE is low, and the memory is addressed sequentially at high speed to generate the TV image. When the controller writes new data to the display RAM, DE goes high, in effect disabling the display addressing.

In addition to clocking the image data out of RAM at high speed, we must also generate TV horizontal and vertical sync pulses if we are to have a stable display. Generation of the short horizontal-sync pulse and longer vertical-sync pulse is accomplished by a dual single-shot (U16) driven by outputs from the addressing counter chain. With a 4-MHz oscillator source, pin 8 of U14 toggles at 61 Hz, providing the vertical-sync timing reference (V). Pin 8 of U12 toggles at 15625 Hz, providing the horizontal-sync reference (H). The TTL high pulses from U16 are combined into a TTL-low composite-sync signal (CS) by Q1. CS is used in the TV output stage to add the sync pulses to the video signal.

A 15.625-kHz horizontal and 61-Hz vertical sync rate are such that the image can be displayed on a standard monitor with only minor adjustment of the vertical and horizontal hold controls. The 4-MHz crystal is readily available for approximately $3, and is an inexpensive approach to generating the TV sync signals. If you want more precise control over the sync frequencies, a 3.93216-MHz crystal can be used in lieu of the 4-MHz unit specified. This yields a vertical-sync frequency of precisely 60 Hz, and a horizontal-sync frequency of 15.36 kHz. (Order this crystal from one of the crystal-supply houses; it costs between $10 and $15—see Appendix I). Use of the more-expensive crystal is probably not justified in any but the most-demanding display applications. The sync rates obtained with the 4-MHz crystal have caused no problems in any of the many display monitors tested, and provides a stable signal for video-taping on most systems. If you plan to route the signal into processing units for broadcast or cable system display, you may want to use the 3.93216-MHz crystal.

If you have a problem with this circuit block, begin by removing all chips from the board *except* for U11, U12, U13, U14, U15, and U16. First, use a logic probe to check for output from the oscillator. Pins 6 and 8 of U11 should be toggling at a very high rate—probably too fast for either the **HIGH** or **LOW** indicators on the probe to register. Your should, however, note a steady flashing of the **PULSE** indicator on your probe. A similar reading should be obtained from pin 11 of U11. If a frequency counter or short-wave receiver is available, you should be able to confirm that the circuit is functioning at 4 MHz. If the circuit is not oscillating, try briefly touching the junction of the two 470-ohm resistors with a grounded test lead. If the circuit oscillates at this point, you probably have a sluggish crystal. There are three tricks that might still salvage the crystal at this point:

1) Try swapping the 74LS00 chip with another one elsewhere in the circuit.
2) Replace the 74LS00 with a standard 7400.
3) Solder a 4700-ohm resistor from the junction of the two 470-ohm resistors to +5 V.

If none of these approaches work, you need a new crystal.

Now use the logic probe to check the outputs, in sequence (pins 3, 4, 5, 6, 11, 10, 9, and 8) of U12 and U14. All should be toggling and, as you move down the sequence, the probes **HIGH** and **LOW** indicators should get brighter since the frequency is steadily decreasing as we move down the counter chain. If you stop getting an indication of toggling at some output of one of the chips, that chip is probably defective.

Perform a similar test on the sequence of outputs from U13 and U15. With DE floating, no outputs should be noted. With DE grounded, the results should parallel those you noted for the counter chain.

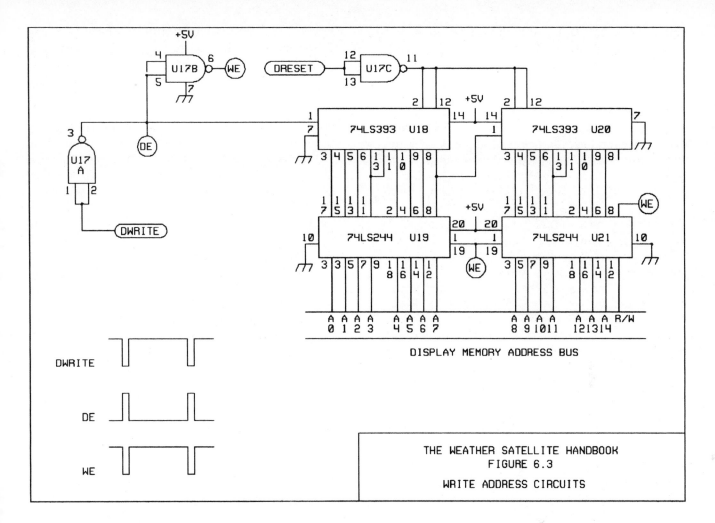

Figure 6.3—Write-address circuits.

Write Addressing (Figure 6.3)

A completely independent write-addressing counter chain, made up of U18 and U20, is used to generate RAM addresses when the controller writes new data to the display RAM chip. When the controller writes a new data byte to RAM, the DWRITE strobe GOES low. This low is inverted to a high by U17A to drive the counter chain. Because this pulse goes high during a data-write operation, it's also used as the display-enable (DE) strobe to disable the normal display addressing. This high pulse is inverted by U17B providing the low write-enable (WE) strobe that controls the two bus drivers (U19 and U21), which control switching of the write-address counter outputs. Thus, during a write operation, the display addresses are disabled, the write-address counters are enabled, and the write occurs to the address specified by the write-addressing

counter chain. When the write operation is complete, DWRITE goes high, the write-address lines are disabled, and the normal display address counters regain control of chip addressing. Because DE goes from a high to a low at this point, the write-address counters also advance by one count because the 74LS393 counter chips transition on a high to low shift at the input.

Not only do these circuits automatically handle the address changes with each write operation, they also ensure that the read/write line (R/W) of the RAM chip is properly shifted to a low during the write operation. Because only 15 lines are required to address display RAM, the last input of U21 (pin 8) is not needed for an address line. The WE strobe, which goes low during a write operation, is routed to the display RAM R/W line, causing the data byte from the controller address bus to be written to the RAM location specified by the state of the write-address counter chain.

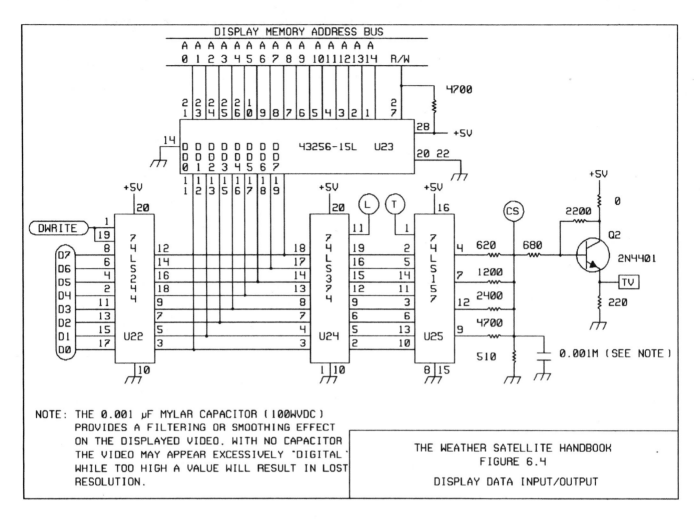

NOTE: THE 0.001 µF MYLAR CAPACITOR (100WVDC) PROVIDES A FILTERING OR SMOOTHING EFFECT ON THE DISPLAYED VIDEO. WITH NO CAPACITOR THE VIDEO MAY APPEAR EXCESSIVELY "DIGITAL" WHILE TOO HIGH A VALUE WILL RESULT IN LOST RESOLUTION.

THE WEATHER SATELLITE HANDBOOK
FIGURE 6.4

DISPLAY DATA INPUT/OUTPUT

Figure 6.4—Display data input/output.

As data for a new image arrives, the controller sequentially writes it to the display RAM. If we are to display the image properly, the display counter chain must be reset to zero at the start of a new image so that the new image data begins at the first location in RAM. This is accomplished by the DRESET strobe from the controller, which is pulled low at the start of a new image. This low is inverted to a high by U17C, resetting the counters.

When troubleshooting, begin by removing all board chips except those shown in Figure 6.3 and U11. Use a clip lead to connect pin 8 of U11 to pin 1 or 2 of U17. Proceed to check out the counters and bus drivers just as described in the troubleshooting section of the Address Circuits.

Display Data Input/Output (Figure 6.4)

The previous discussion has shown how a write operation (DWRITE goes low) controls the proper addressing of the display RAM. This same strobe also controls the interconnection of the controller data

bus (D0-D7) and the display-RAM data bus (DD0-DD7 of U23). When DWRITE goes low, the data on D0-D7 is gated through another bus driver (U22) and is thus applied to the RAM DD0-DD7 data lines. The RAM R/W line is normally pulled high during display by a 4700-ohm pull-up resistor but, as noted in the previous section, R/W goes low during a write operation. Thus, when DWRITE goes low during a write from the controller, the write-address counters are enabled, the controller data are routed through U22 to the RAM data bus, and the RAM R/W line is pulled low to write the data to the memory chip. At all other times, the controller data bus is isolated from the RAM data bus and the R/W line is high.

During normal display addressing, the chip is being sequentially addressed at high speed, and the data on the RAM data bus (DD0-DD7) is the stored data we want to convert to a TV image. It takes a finite amount of time for the new data values to stabilize with each address transition. The address cycling is occurring so fast that, much of the time, the output data is invalid

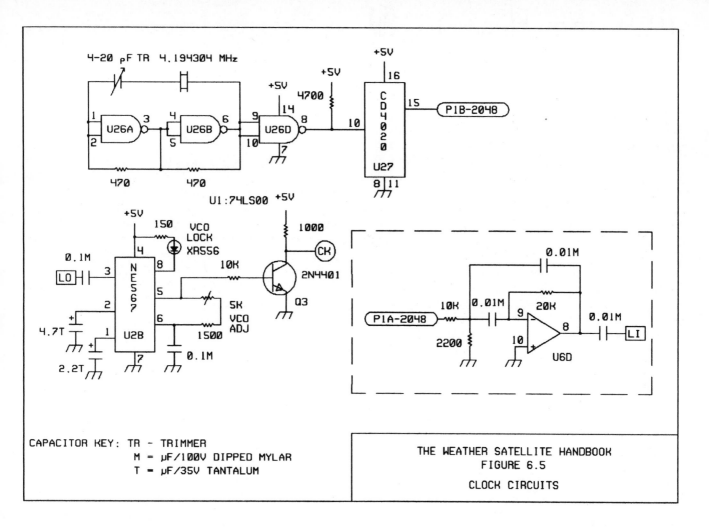

Figure 6.5—Clock circuits.

because the values are in transition from the value for the previous address and the value for the new address. An octal latch (U24) is used to provide a snapshot of the RAM data bus to get around the problem of transient data. The latch control strobe (L) is generated by the display-address circuits at a stable point in the address cycle. This strobe freezes the RAM data values on the output of U24.

The 8-bit data at the output of U24 actually represents a pair of 4-bit pixels, formatted as follows:

Data Bit	Pixel	Bit
7	pixel #1 – bit 3 (MSB)	
6	pixel #1 – bit 2	
5	pixel #1 – bit 1	
4	pixel #1 – bit Ø (LSB)	
3	pixel #2 – bit 3 (MSB)	
2	pixel #2 – bit 2	
1	pixel #2 – bit 1	
Ø	pixel #2 – bit Ø (LSB)	

The two 4-bit pixels must be properly sorted out to provide the TV display. This sorting function is provided by U25, a data demultiplexer. Essentially, U25 functions as a 4-pole, double-throw switch, controlled by the toggle strobe (T) from the display address circuits. U25 is wired so that pixel #1 data are routed to the four output lines of U25 during the first half of each byte address cycle. During the second half of each address cycle, the pixel #2 data are routed to the output. The four output data bits from U25 are applied to a weighted resistor network in the base circuit of Q2. This transistor sums the digital data values, generating an analog voltage swing on its emitter, producing the analog TV signal. The composite sync (CS) signal is also applied to the base of Q2, mixing the sync signal with the analog video to provide a complete TV video signal.

If gray-scale loading to the display is erratic, the problem is probably localized around U22. If the display data are unstable, U24 is one place to look for a problem. Improper sequencing of gray-scale steps probably is a function of misplaced or improper resistors in the output of U25.

Clock Circuits (Figure 6.5)

The final function provided by the display board is the generation of a precision clock signal that the controller uses to time image sampling and other time-critical functions. The heart of the clock circuit is a 4.194304-MHz oscillator (U26A and U26B), with output buffered by U26D. The 4.194-MHz signal is routed through a CMOS counter chip (U27) wired to provide a divide-by-2048 function. The output of U27 is a 2048-Hz signal. This 2048-Hz clock is ideal for timing image sampling since there are precisely 512 clock transitions in a 250-ms video line (240 LPM) or 1024 in a 500-ms line (120 LPM). This 2048-Hz signal must be set with extreme accuracy, so a trimmer capacitor in the 4.194-MHz crystal circuit is used to set the oscillator precisely to the marked crystal frequency.

Although direct use of the clock signal would suffice for displaying live pictures coming directly from the receiver, it would not serve for display of recorded images. No matter how high the quality of an analog tape audio system, there are inevitable speed changes in the mechanical tape system during record and playback that would alter critical system timing. We avoid this problem by recording the 2048-Hz clock signal on the left channel of the tape deck, while the satellite video signal is recorded on the right channel. Since the output of U27 is a TTL square-wave, it is not ideal for audio recording. The square-wave signal is converted to a sine wave by routing the signal (2048) back to the controller board where a section of U6 (U6D) is used as an active bandpass filter. The output of U6D is routed to the left channel input (LI) of the tape deck.

On playback, the recorded clock signal on the left channel is applied to the input (LO) of a phase-locked tone decoder (U28). U28 has an internal voltage-controlled oscillator (VCO), controlled by the VCO **ADJ** control. When the VCO control is set so the VCO free-runs near 2048 Hz, the VCO locks to the taped 2048-Hz signal and tracks any frequency variations induced by speed changes during recording or playback. The internal VCO signal is buffered by Q3, providing the system clock signal (CK). Because CK tracks the original recorded tone, all system operations are as accurate using the recorded signal as they might be for a live reception. If a recorder is not used, an audio cable is used to jumper the **LI** and **LO** jacks on the rear apron of the scan converter, locking U28 directly to the crystal-controlled output from U6D.

Follow the procedure outlined under the discussion in the Display-Address Circuits to determine whether the oscillator (U26) is functioning. A logic probe should indicate a steady pulse train at pin 15 of U27.

If the LED indicator does not light with 2048-Hz input, use a clip lead to ground pin 8 of U28. If the LED does not come on, the LED is defective. If the LED does light, the problem is probably with U28. A logic probe should also show a steady pulse train at the collector of Q3.

Construction

The construction considerations for the display board parallel those of the microcontroller. Construction is quite easy using the circuit board available from Metsat (see listing at the end of this chapter). The board is 8.9 × 6.25 inches in size, double-sided and has plated through-holes. Figure 6.6 shows one of these boards fully wired. Simply use sockets for all ICs, insert all devices as shown in the exploded component layouts that accompany the board, and double-check orientation of IC sockets, tantalum capacitors, and diodes prior to soldering the parts into place.

If you are going to wire the board from scratch, follow the same procedures you used when doing the controller board: heavy-gauge power connections, plenty of bypass capacitors, and the use of schematic copies and colored pencils to check off and later verify wiring. The only critical areas are the two crystal oscillator circuits where short leads should be used. Considerable latitude is possible in routing all the other leads although, in general, you should keep the leads in the display-address circuits shorter than those in the write-address circuits.

CHECKING OUT THE BOARDS

Both the controller and display boards should be checked out on the bench prior to wiring everything into a cabinet. You'll need a supply of clip leads, a 12-V power supply and a source of 5 V. For the latter, temporarily connect your 12-V supply to a 7805 regulator chip bolted to your cabinet (to provide a heat sink). Be sure there are no loose parts on the bench that might cause a short circuit beneath the board(s) you are testing. In all of the steps that follow, power-supply connections imply that you connect a positive power lead between the board and the supply as well as a lead from the foil ground plane (around the outer edge of the board) and the negative supply.

Insert all chips in the display board *except* the RAM chip. Connect a TV monitor to the TV output and ground and connect the board to the 5-V power source. The TV screen should be white, and you should be able to lock a stable display with a slight adjustment of the **HORIZONTAL** and **VERTICAL HOLD** controls. If the display is not white, check the Display Data Input/Output circuits. If there is no sync—or sync is erratic—check the display-address circuits.

Power down, insert the RAM chip, and power up. You should still have a stable TV display but now—in-

Figure 6.6—A completed display board. The clock oscillator and associated 4.194-MHz crystal are located in the upper-right-hand corner; immediately to the left is the clock divider. The next three chips to the left are the write-address control and counter circuits, with their associated 74LS244 bus drivers immediately below. The display-address clock circuit and 4-MHz crystal are just to the left of top center, followed (to the left) by the display-address counters and their associated bus drivers immediately below. Along the right edge of the board, immediately below the clock oscillator, are the diodes for the front-panel switching circuits; just below the diodes is the vacant socket (P1A) that provides the interface connection to the microcontroller. Below P1A and to its left is the bus driver for the switching circuits, with the write-address buffer immediately to the left. Below this pair of chips is the NE567 tone decoder used in the phase-locked clock circuit, the VCO lock indicator LED, and the VCO potentiometer. The 43256 32-kbyte × 8-bit video display RAM chip is the large device in the lower center. Immediately to the left of the RAM chip is the display data latch, with the data multiplexer to its left. Above the data multiplexer is the dual single-shot that generates the TV-display sync pulses.

stead of a white screen—you'll have a somewhat chaotic display of patterns representing the random contents of the display RAM at power up. The pattern varies from RAM chip to RAM chip but, whatever the pattern, it should be constant. Power down.

Assuming you have a programmed EPROM for the controller board, connect the controller to a 12-V source and interconnect the controller and display boards with a 16-conductor header cable. Observe the color pattern of the cable and the key ends of the P1A and P1B sockets to assure the proper polarity in interconnecting the two boards. Apply power to both boards. For about one second, you'll see the previously mentioned random pattern on the screen, then it will be replaced by the display METSAT 1700 in black letters against a white banner across the center of the screen with a gray-scale pattern above and below the banner (Figure 6.8).

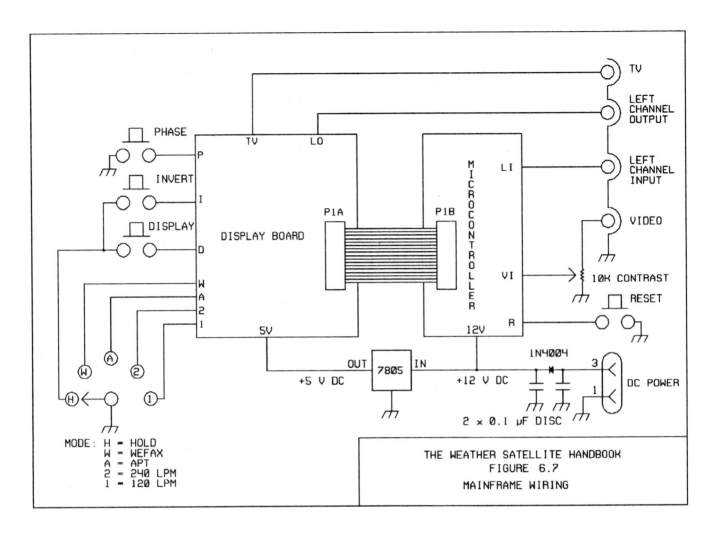

Figure 6.7—Mainframe wiring.

The gray-scale consists of two cycles of eight steps, ranging from black at the left to white on the right. Because of normal monitor overscan, the darkest steps on the left will probably be out of view, as will the lightest steps on the right. Because the pattern is repeated twice across the screen, the center of the screen area should have examples of the darkest and lightest steps. Adjust monitor brightness and contrast controls to provide a bright display that clearly differentiates each step from its nearest neighbors. The darkest bar should be pure black, while the brightest bar should be a clean white, with no smearing or trace distortion (indicative of too high a brightness setting). If the steps are not in a clear sequence, or one or more of the bars appear to be twice as wide as the others, mis-wiring of the output resistors at U25 is indicated.

With power still applied, use a clip lead to jumper point **LO** on the display board and **LI** on the controller. Adjust the VCO control through its range; the

LOCK LED near U28 should come on. Leave the control set at the midpoint of the range in which the LED indicator stays on. Disconnect the clip lead between **LO** and **LI**. An amplifier or pair of headphones connected to point **LI** and ground should produce a clear audio tone. Power down.

Connect one side of your 10-kilohm **CONTRAST** pot to ground. Connect the other side to **LI** and the middle lug to **VI**. Adjust the control so the center arm is at the ground end, and apply power. Once the gray-scale has been posted, connect a clip lead between switch point 2 on the display board and ground. After about two seconds, short, transient black streaks should be visible on the gray-scale display. After several seconds, a zone of black should begin to move downward—line by line—from the top of the screen. If nothing happens, check the Display Microcontroller Interface circuits.

Steadily increase the setting on the **CONTRAST** control; the trace should begin to brighten. When the trace finally becomes white, rotate the control slowly back in the other direction; the trace should darken to black. If there is no change in trace brightness, or the display patterns do not show a smooth gradation, troubleshoot the video circuits. Disconnect the clip lead from switch point 2 and the writing to the screen should stop.

Briefly touch the free end of the grounded clip lead to switch point **D** on the display board: A random pattern, representing the random contents of the controller RAM, should be written to the board, replacing the previous display. Briefly touch the switch lead to point **I**, and a similar pattern should be written to the display, but it will take two to three times longer than the previous case. If you go back and forth between these two tests and carefully examine the screen, you'll note that the pattern obtained when I is touched is an upside-down version of the pattern obtained when touching switch point **D**.

Finally, touch the grounded clip lead to point **R** on the controller board. After about one second, the banner and gray-scale should overwrite the previous display.

Congratulations! You now have a completely functional set of boards! The only thing between you and a completed scan converter is getting them packaged in a cabinet.

MAINFRAME WIRING

The scan converter Mainframe Wiring is summarized in Figure 6.7. In the case of the Metsat Model 1700 commercial version of this circuit, both boards are packaged in an LMB 11 × 4 × 7-inch cabinet. This display board is mounted in the bottom of this cabinet with the controller board mounted near the top, supported by 1- × 1-inch aluminum angle brackets. The rear apron contains the **POWER** connector, the **RESET** push-button switch, and the **TV**, **LI**, **LO**, and **VIDEO** phono jacks. The front panel has the rotary **MODE** switch, the **CONTRAST** control, and the **DISPLAY**, **INVERT**, and **PHASE** push-button switches. A **POWER** switch and associated LED indicator are optional; it's just as easy to control the scan converter by using the 12-V dc power-supply **ON/OFF** switch. Complete operating instructions for the scan converter are included in Chapter 9.

SCAN-CONVERTER COMPONENTS

A complete listing of components for the controller, display board, and mainframe wiring is included

Figure 6.8—Initial screen display when the microcontroller card has initialized the display board.

in Appendix II. All of these components are available from parts vendors listed in Appendix I.

Metsat Products (see Appendix I) supplies the following items to assist in your scan converter project:

- Scan-Converter PC-Board Set. Includes the controller and display circuit boards as well as the A-bus interface circuit described in Chapter 7. All boards are double-sided with plated through-holes, and are made to the most exacting standards. The set comes complete with parts layouts and documentation to supplement the descriptions provided here. Order the Model 1700 Board Set. The price is $100, postpaid in the US and Canada.
- Model 1700 ROM. Programmed EPROM for the microcontroller of Chapter 5. It provides all of the basic functions described in Chapter 9, and all of the interface functions described in Chapter 7. It's priced at $50 postpaid in the US and Canada. (All prices are subject to change without notice.)
- Model 1700 Scan Converter. Wired and tested scan converters are available. Contact Metsat for current pricing and delivery information.

Chapter 7

Scan-Converter and Computer Interfacing

INTRODUCTION

One of the more-powerful features built into the scan converter is its ability to talk to an external computer via the two parallel ports on the microcontroller card. Depending on the sophistication of your computer system, applications can range from simply using the computer for image storage and retrieval, to more complex operations such as image processing and high-resolution image display.

In order to take advantage of these options, you must develop a suitable hardware interface between your computer and the scan converter, and use a software package that lets the two communicate predictably. Central to both of these, of course, is the computer itself. I'll digress in the first section to discuss a few aspects of computer selection, followed by some notes on the suppression of the inevitable RF interference that the computer generates. When both of those topics have been covered, we'll look at interfacing and software questions. The final section of the chapter deals with some of the basic functions you can implement with the scan converter/computer connection.

SOME NOTES ON COMPUTERS

If you intend to purchase a computer for use with your satellite station, a number of factors will impact your choice. Based on pure cost-effectiveness, it is hard to go wrong with an IBM PC or one of its clones. Add-in cards will be required for most of our applications, and the sheer size of the PC market results in a situation in which you have the widest possible range of available functions at the lowest possible price. Sheer speed or the elegance of the operating system have little bearing on the overall choice. What we want is power, memory, and—overall—flexibility.

The reason for bypassing other computer systems is strictly practical. Eight-bit systems such as the Color Computer, Apple II, and Commodore will all do just fine

for basic image saves—and even video processing—when used in conjunction with the scan converter. If you have such a system, by all means put it to use. Unfortunately, there is no simple path to upgrading these computers to more advanced options, so I cannot recommend them as start-up systems on which to build.

The Macintosh line offers superb computing power and a very intuitive user interface. Unfortunately, the systems are very expensive in terms of what's needed for use in a satellite station, and upgrade costs are equally high. If you have a Mac (or other reasons for acquiring one), then use it. The Mac is not, however, the most cost-effective way to build a foundation for satellite work.

The Amiga is a very fine system with superb graphics and an interesting range of I/O options. Unfortunately, Amigas operate in their own software and hardware universe. Again, you can use it for satellite work, but unless you have other applications in mind, such a system is not the best starting point.

A base-line system can be built around an 8- or 10-MHz 8088 XT clone. An 80286-based computer provides even more options (without a great price differential) in terms of speed and the ability to address megabytes of RAM. If you're configuring the system from the start, go for the full 640 kbytes of RAM, a 360-kbyte, 5¼ inch drive (usually standard), and a second 1.44-Mbyte, 3½-inch drive. The latter is very useful because it lets you save a larger number of files per disk. When configuring a system, you should take care in the selection of a disk controller by looking to the future. You may not be prepared to add a 3½-inch drive or a hard drive at the outset, but if you select a controller with enough flexibility, adding new drives can be simple and economical. If you start with a limited-ability controller, replacement of the controller may be required if you want to add additional drives.

Computer display options are really quite limited. If you won't be using the computer for image display, any video card will do: A CGA-compatible card with composite output and a green or amber monitor is probably the most cost-effective approach. Such a system can be used to store and manipulate images, but all data would be passed to the scan converter for actual display. This is a perfectly sound way to start.

Once you decide that you want to use the computer for display, however, choices are limited. Using a CGA display for satellite images is basically a waste of time. The scan converter provides a much better image. A CGA display can be used for false-color display, but that's a very limited application.

The next step up, EGA, represents a significant cost increase because it's basically a color display system. You'll need an EGA display card and an EGA monitor. Although the spatial resolution of the EGA system is superior to that of the scan converter, it's questionable whether the improvement is worth the added cost, and you are linked primarily to false-color display. If you already have an EGA system, you can put it to good use in conjunction with the scan converter, but I would never purchase an EGA display strictly for satellite use!

The display system of choice for the PC line is VGA. VGA cards can be obtained for $250 or less. VGA monitors come in two flavors—white monochrome displays that provide a 64-level gray-scale display, and color monitors that display either monochrome or color, depending on the software palette selection. A VGA monochrome monitor is excellent for displaying satellite images. Such monitors can be obtained for as little as $150. Almost all of the images in this book were displayed on such a system. If you are configuring a computer from scratch, start with monochrome VGA. A VGA card and monitor can be added at a later date, but if you do upgrade from another display system, you'll have to remove the other display card because both graphics adapters cannot coexist in the same system. That's why, in the long run, it is slightly less expensive to use VGA from the start, assuming you can handle the initial cash outlay.

Building a computer from components is a simple task that takes only a few hours. Based on today's prices, you could configure your satellite-system computer as follows:

Case with 200-W power supply	$190
10-MHz 80286 motherboard	200
1 Mbyte of RAM	126
Floppy controller (4 drives)	50
360-kbyte, 5¼-inch drive	70
1.44-Mbyte, 3½-inch drive	100
Scan-converter interface	100
16-bit VGA card	250
14-inch, white, VGA monochrome monitor	120
Total	$1206

This kind of system has plenty of room for expansion (including RAM and hard disk), but is a powerful system as it stands. The prices are realistic, having been taken from a recent issue of *Byte*. The options quoted are for quality components. I could have cited lower prices, but they would have represented boards of unknown quality. By omitting the 3½-inch drive, using an 8088 motherboard, an 8-bit VGA card and 640 kbytes of RAM, you can drop the system price to the $900 range and still have the display quality equivalent to the images in this book. Given the concerted rush to high-end PC systems, you can often pick up a basic XT at fire-sale prices and upgrade it at quite reasonable cost.

I am old-fashioned enough to consider a thousand dollars to be "real money," not an amount to be lightly allocated from one's family budget. The whole key to the scan-converter concept involves incremental expansion. A $250 scan converter will function standalone. If you have an existing computer, it can be interfaced immediately and provide the tangible benefits of being able to save images to disk, process video data, and perform other functions. Gradual computer-system upgrades won't deal a paralyzing blow to the family finances.

LIVING WITH A COMPUTER

A computer interface to your scan converter, or one of the many computer-based display systems, can provide you with tremendous power in manipulating satellite images, but creates new problems in terms of RF interference. Ironically, the newer, more-powerful computers, are more of a problem than some of the earlier eight-bit home computer systems. The newer units have higher clock frequencies; harmonics and spurious products will be stronger (everything else being equal) than was the case with the earlier systems. Although interference on S-band (for GOES reception) is rarely a problem, reception in the 136- to 138-MHz polar-orbiter-satellite band can be severely effected. If you can hear noise from the computer on an unoccupied satellite frequency, it will certainly have some effect on signal reception, particularly early and late in a pass when the satellite's signal is weak.

Generally, computer manufacturers go to fairly great lengths to shield their systems, if only to pass the required FCC certification tests for RF emissions. There are two general levels of FCC classification for computer devices. Class-A certification is for business use, while class B covers the home environment. Class-A certification is more lenient because the usual business environment does not involve problems with TV or radio interference. Class B standards are stricter because the computer has to limit RF emissions sufficiently to permit the operation of nearby radio and TV receivers, cordless telephones, and the like. Conceivably, a class-A-certified computer could cause interference to your neighbor's radio and TV reception and, if they object, fixing the situation is your problem. Most manufacturers state the computer's FCC certification type, and you should be aware of the potential for increased RF interference if you are using a class-A-certified system.

All of the RF suppression efforts are in vain, however, if we provide an RF pathway for these signals to leave the computer. Let's look at two contrasting cases, both involving interfaces to the scan converter, to see the source of some of the problems.

The MetraByte interface card (one of the interface options I'll discuss in the next section) is located entirely within the computer. The only signal lines that exit the system are the parallel data lines to the scan converter. Much of the time these lines are completely inactive and, even when they are busy, signals are switching at audio—not RF—rates. The result is a relatively low potential for additional RF interference beyond that created by the computer itself. The multi-conductor cable provided by MetraByte is fully shielded, so even miscellaneous signals coupled out on the data lines are reduced greatly in amplitude.

In contrast, the flat ribbon cable that connects to the A-Bus card (another of our interface options) inside the computer carries a buffered version of the computer data bus and many address lines. This cable is unshielded and, whenever the computer is on, these signal lines are switching at RF rates. Although the header cable between the external TTL I/O card and the scan converter is relatively quiet, the ribbon cable from the card back to the computer can be a major source of interference. To minimize this problem, lay a piece of bare bus wire along the length of the ribbon cable, then wrap the cable in one or two layers of household aluminum foil. The bare wire at either end of the cable can then be connected to ground to create an effective RF shield around the cable. In general, the better your station ground, the better the shield. Once everything is set up, the foil can be wrapped in electrical tape so that you won't inadvertently short something to ground.

In general, all cables that exit your computer should be shielded. This is an absolute requirement if the cable is carrying computer address or data signals. Ribbon cables may have to be used to carry such signals in the case of some computers and/or interface options. In such cases, you should shield the cable(s) in question as described earlier.

INTERFACE HARDWARE

The popularity of the many kinds of personal computers has led to a thriving marketplace with vendors supplying interface hardware—usually involving plug-in cards—that allow these computers to perform specialized tasks. The scan converter has two 8-bit parallel ports. One is dedicated as an input port (I∅-I7) to receive data from the external computer while the second is a latched output port (O∅-O7) dedicated to sending data to the external computer. What your computer requires is two similar parallel ports—an input port and a latched output port—a total of 16 I/O lines.

The terminology gets a little mixed up in places because vendors take a few short cuts in labeling their boards. The kind of board we want is usually called a "TTL I/O" board, or something similar. In this application, we need at least 16 bits of parallel I/O. In most cases, the interface board is configured so that the available bits can be configured as either input or output, using software control. In some designs, however, the lines may be dedicated as input or output, in which case we need at least eight input and eight output lines with the latter being *latched* outputs: Once a particular pattern of bits has been written to the output port, that pattern remains at the output until a new pattern is written. Be careful of cards labeled as providing simply one or more "parallel ports." Often what is really meant is "parallel printer ports," a fact you can usually determine by reading additional specifications for the interface hardware. Although it is often possible to modify parallel printer ports to provide the needed functions (some combination of hardware modification and/or software), the task is simpler if we start with a card specifically designed for TTL I/O.

Products from two different vendors are highlighted below. These are certainly not the only options, but I am familiar with them; they are high-quality products that you can use without fear of damaging your computer, and they cover a wide range of computer models. A check of the magazines that cover your particular computer will undoubtedly reveal other sources that may do just as well. The examples provide at least one option for the following computer models:

Apple II and //e
Apple Macintosh II
Apple Macintosh Plus
Apple Macintosh SE
Commodore C64 and 128
IBM PC
IBM PC/XT
IBM AT
IBM PS/2 Model 25
IBM PS/2 Model 30
IBM PS/2 Model 50-80 (Microchannel)
Radio Shack Color Computer (all models)
Radio Shack TRS-80 Model 3, 4, and 4D

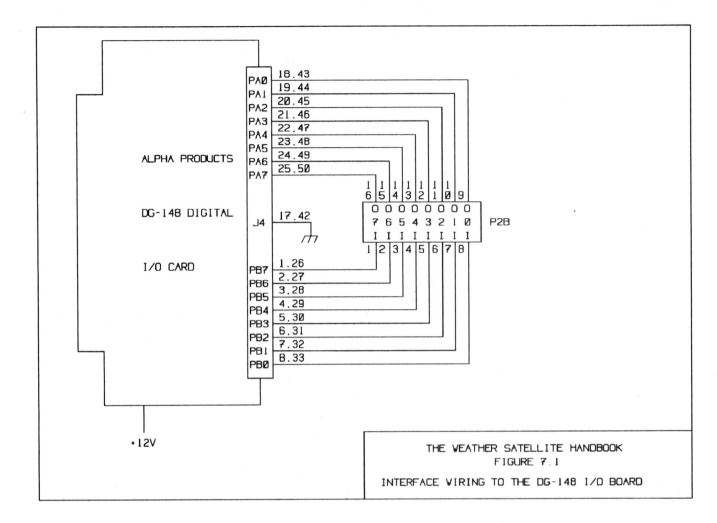

Figure 7.1—DG-148 card interface.

What you can expect to pay for interface hardware is a function of the cost of your computer and the complexity of its bus. Hardware for relatively simple computers is usually not expensive. The same is true for the various varieties of the IBM PC and their many clones because astronomical numbers of these computers exist and vendors are competing for customers. More-expensive computers, especially those with a proprietary or complex bus (IBM PS/2 Microchannel systems and Apple Macintosh, for example), command a premium because the market for add-ons for these machines is more limited and the hardware prices reflect this.

A-Bus System

Alpha Products markets an innovative system called the A-Bus, which is designed to provide a wide range of I/O functions for a number of personal computers. Creating our interface using the A-bus system requires three components:

1) An adapter card that plugs into the bus of your computer

2) A 24-line TTL I/O card (DG-148, $72)

3) A ribbon cable to connect parts 1 and 2 (CA-163, $24)

It is the adapter card that is specific to your computer. At present, Alpha markets the following adapter cards:

AR-133 (PC/XT/AT/PS/2 Model 25 and 30) $69
AR-134 (Apple II, //e) $52
AR-139 (Commodore C64 and 128) $48
AR-138 (Radio Shack Color Computer) $49
AR-132 (Radio Shack TRS-80 Model 3, 4, 4D) $54

As this edition goes to press, Alpha Products has announced a new interface card designed specifically for the Microchannel bus, but I have not seen a price on this unit.

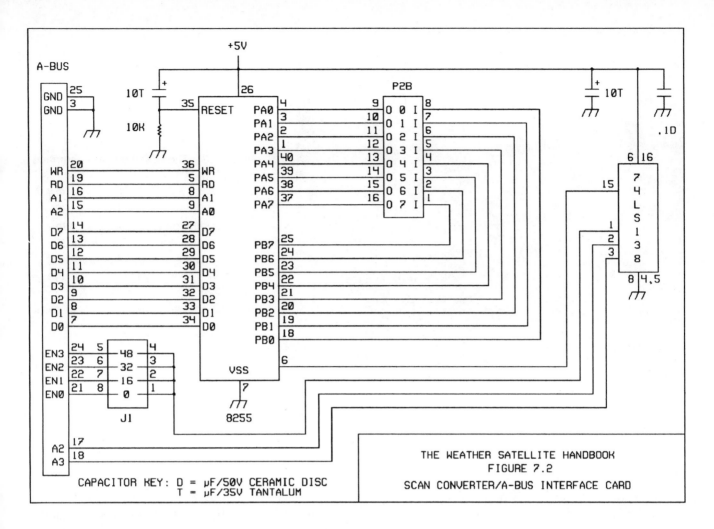

Figure 7.2—Scan-converter interface for the A-Bus system.

Installation, address decoding, and initialization software are covered in the documentation package. The DG-148 TTL I/O board provides three 8-bit ports (A, B, C). In our application, we want Port A set up as an input, connected to the output port of the scan converter, while Port B is used as an output, connected to the input port of the scan converter. If a 16-pin DIP socket is wired to the DG-148 board as shown in Figure 7.1, a 16-conductor ribbon cable with header plugs at each end can be used to connect P2 on the scan converter microcontroller with the DIP socket that is wired to the DG-148 board. You'll also require a ground connection between the computer and the scan converter and a +12-V power conductor.

I'm using the A-bus system and its performance has always been flawless. If I had to cite a single problem with this approach (other than more stringent shielding to quiet the ribbon cable) it would be the price of the DG-148 video card. This is a versatile card, but it is expensive for the handful of chips used, not to mention the tedium of wiring the DIP socket to mate with the scan converter. The Metsat PC board set for the scan converter now includes a small I/O card that is designed to replace the DG-148 card and has two major advantages. The board plugs right into the A-bus ribbon cable from the adapter card for your system, but the total cost for components is less: somewhere between $10 and $15. The second advantage is that the card is designed for use with the scan converter, so the required DIP socket is on the board. The circuit for this card is shown in Figure 7.2, with a photograph of the card in Figure 7.3.

An 8255 I/O chip is used to produce three 8-bit parallel ports (24 bits of TTL I/O) in which any one of the bits can be programmed for input or output. In this application, port A (PA0-PA7) is used as an input to read the outputs from the scan converter. Port B (PB0-PB7) is used as a latched output to drive the scan converter input port. Port C (PC0-PC7) is not shown

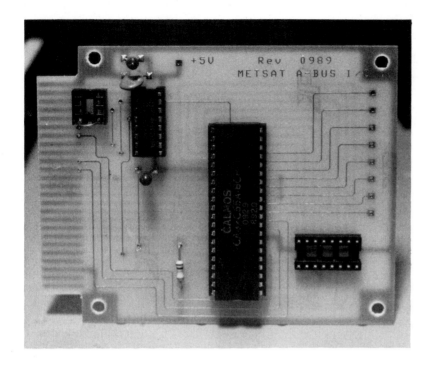

Figure 7.3—The scan-converter interface card for the A-Bus system. The large device in the center of the board is the 8255 interface chip with the 74LS138 decoder immediately above and to the left. The small, 8-pin DIP socket is J1, which provides a range of port-address selection. The empty DIP socket in the lower right is for the header cable from the scan-converter microcontroller board.

on the diagram because it is not used for scan-converter interfacing. But eight pads connected to this port are available on the card for any use that might occur to you.

Address decoding is provided by a 74LS138; jumper block J1 provides a selection of addresses which operate in conjunction with the address selection on your computer adapter card. With your adapter card jumpered for a specific base address, you can use J1 to select a base address for the card that is 0, 16, 32, or 48 above the basic card address. Let's assume you have a PC adapter card that has the factory-installed jumpers for a base port address of 512 decimal (200h). If the jumper at J1 on the interface card were set up between pins 1 and 8, the following port addresses would be implemented:

Card Port	Decimal Address	Hex Address
A	512	200h
B	513	201h
C	514	202h
Control	515	203h

In order to set Port A as an input and B as an output, we need to write a 144 (decimal) to the interface control port at address 515d (203h). This can be accomplished in BASIC with a simple output statement:

OUT 515,144

From this point on, the port is properly configured for use with the scan converter. Data values from the scan converter can be read with a simple input command from Port A:

X = INP(512)

where X represents the data value coming from the scan converter. Assuming we had a variable (let's call it X) that we wanted to send to the scan converter input, we could do it with a simple I/O output command to port B:

OUT 513, X

Connecting the I/O card couldn't be simpler. Simply insert the card into the A-Bus cable coming from the adapter card in your computer, connect a 16-conductor header cable between P2B on the interface and P2 on the microcontroller card, and supply a +5-V power line and ground return. The adapter can be external to the scan converter, or it can be mounted inside with the edge connector protruding from one side to accept the connector on the A-Bus cable.

The only precaution you must observe is that power be applied to the scan converter/interface card prior to booting your interface software. This has nothing to do with the possibility of damaging anything, but is simply because you cannot initialize the interface card without power applied!

MetraByte Corporation

MetraByte Corporation markets a line of high-end interface boards for the complete IBM computer line (including Microchannel PS/2 Models 50-80) and most Apple Macintosh models. For the IBM PC, XT, AT, and PS/2 Models 25 and 30, you can use the PIO-12 board ($120). For the Microchannel PS/2 Models 50-80, the microCPIO-12 ($299) will do the job. Both are plug-in cards where connections are made via a 37-pin connector at the rear slot cut-out. Both boards use the same 8255 chip used in the Alpha DG-148 board (and the interface card we just discussed) and provide the same three 8-bit ports (A, B, C). Interfacing is simply a matter of wiring the 16-pin DIP socket to a connector to mate with the MetraByte cable from their board. Complete documentation is provided and this, in conjunction with Figure 7.1 and 7.2 should provide all the information you need.

The Macintosh II has NuBus slots that accept their MACPIO-24 ($200), which also provides three 8-bit programmable ports. The Macintosh II, Plus, and SE models can also be interfaced via their SCSI ports. The MetraByte SCSI8DIO ($490) is an external unit that connects to the SCSI port and provides an 8-bit dedicated input port and an 8-bit dedicated output port.

Other options may vary in detail, but if the input and output ports of your interface are wired as shown in Figure 7.1 and 7.2, you should encounter few problems. Just remember to always provide a ground connection between the computer and scan converter since the 16-conductor header cable has no provisions for the needed ground return.

SOFTWARE

The ROM-based software of the 1700 scan converter supports three major categories of functions involving an external computer. Arranged in order of complexity, these include:

1) Control of 1700 operation modes/functions
2) Transfer of 256 × 256 (32-kbyte) images between the 1700 and the scan converter
3) Streaming of high-resolution image data in real-time from the 1700 to the external computer

The first two options are what we might refer to as "command" options because the scan converter executes specific functions when requested by the computer. The streaming option (item 3) is fundamentally different in that the scan converter does this without being asked; we'll look at the process a bit later.

Several thousand times each second, the scan converter looks to determine what you want it to do. Each look consists of two operations. First, the front-panel interface is scanned to see if any of the switches have been activated. If so, the scan converter begins to perform the function requested, be it loading a new image, or one of the display options. If it initiates an operation as a result of your activating one of the front-panel switches, it continues to check the switches to determine when to phase an image, stop display, etc. Responding to switch inputs is the primary operating mode and the basis of the stand-alone capabilities of the scan converter.

If the scan converter detects no front-panel-switch activity, it then takes a look at its input lines (the output from your computer) to determine if an external request has been generated. If it detects a meaningful request (based on the routines programmed into its ROM), the scan converter performs that task. While doing so, it will, in most cases, check the port constantly to be sure that you haven't issued a new command, such as an abort. When executing a function requested by a controlling computer, the scan converter ignores its front-panel switches and responds only to commands coming through the computer port. It's this ability to respond to specific requests that's the principle reason for interfacing even a simple personal computer system to the scan converter.

In order to demonstrate how to employ the potential of the scan converter computer I/O port, I wrote a program called WSH1700.BAS, which is presented in Appendix III. This program is divided into a number of modules for performing various tasks and is liberally commented. In discussing the various functions, I'll be referring to specific modules in this listing. The program represents a complete entry-level program written entirely in BASIC. Although this results in a speed penalty in terms of the image-transfer functions, the program can be adapted to run on virtually any personal computer equipped with BASIC. The program is also available on the PC compatible *Weather Satellite Handbook Program Disk* available from ARRL. (See Appendix I.) The version of the program on the PC disk is compiled, which increases its speed of execution. In addition, the image-transfer routines in the compiled program are implemented in assembly language, making them extremely fast compared with the raw BASIC routines.

Each version of the program is completely menu-driven, with all functions available from the **MAIN MENU** by the use of single keystrokes. In order to make program operation as simple as possible, the program assumes that your interface card (A-Bus or Metrabyte) is set up to address port A as input at 512 (200h), and port B is used as input at address 513 (201h). Port values are assigned in Module 01 (lines 7Ø-13Ø). Since these are the factory default addresses for both cards and nothing else (other than the Game Card) uses these addresses in virtually all systems, they should be trouble-free. Line 12Ø does the actual port initialization and 13Ø sets all output bits high (255) in the standby configuration.

Mode/Function Requests

When I first designed the scan converter, I had in mind only the image-transfer and streaming functions. With the first dozen or so units out and in use, I began to get some interesting suggestions for additional options. This whole category of functions owes its existence to a very simple question asked by Owen Scotland down in the sunny Cayman Islands. Owen really liked his scan converter and he had it interfaced to a PC. His question was really quite simple: With the system interfaced to the computer, why should he have to continue to use the front-panel switches? Shouldn't it be possible to program the 1700 ROM so it could enable the various modes and functions of the scan converter from the computer keyboard? Of course it was possible, but I hadn't thought about it! Once I did think about the potential, the results went beyond what Owen had actually asked for, but no one is complaining!

Display Functions

The front-panel switches provide for three basic display functions: DISPLAY, which simply causes the scan converter to pass the image in its RAM to the display screen; INVERT, which allows the display of an inverted version of the image in RAM; and COMPLE-MENT, where the image in RAM is passed to the display in complemented form by pressing the PHASE switch when the MODE switch is in the HOLD, or standby, position. These functions can be implemented by the external computer by simply writing the proper value (MODE) to the OUTPORT using the OUT I/O statement in BASIC. The legitimate values for the MODE variable are as follows:

DISPLAY	2
INVERT	1
COMPLEMENT	7

These operations are called from the **DISPLAY CONTROL** portion of the **MAIN MENU**, with the actual work performed by Module 05 (lines 115Ø-12ØØ) in the program listing. Line 117Ø performs the actual request for this display function. We have to do just a bit more than simply write the value to the port how-

ever, since if we were to just write it and leave it there, it would be equivalent, for example, to pressing and holding the DISPLAY switch. If we did that, the scan converter would continuously re-display the image. That's because each time it completed the display, it would read the switches again and see another DISPLAY request! We want to place the request on the output port long enough for the computer to recognize and act on the request, but the port should then be returned to standby (255) so the scan converter won't repeat the action unnecessarily. Given the high speed of the scan converter, relative to BASIC, you could probably write the MODE value to the port and then write a 255 with the next statement. Because it takes the scan converter about a second to perform any one of these tasks however, it is sufficient to return the port to a value of 255 before the scan converter is finished. Line 118Ø provides a very short time delay, and line 119Ø returns the port to the standby mode.

Operating Modes

In order to keep front-panel switching simple, the scan converter supports four basic imaging formats with the front panel MODE switch:

WEFAX	automatic display of WEFAX images
APT	manually phased visible or IR NOAA display
240 LPM	manually phased WEFAX or COSMOS display
120 LPM	manually phased Meteor, or side-by-side visible and IR NOAA display

This combination provides the maximum flexibility to handle any image format with minimal complexity in the switching network. It's possible, however, to design automatic-phasing routines for NOAA visible, NOAA IR, and standard 120-LPM Meteor images that function with a high degree of reliability—provided you have an essentially noise-free signal. Such routines are quite specific, however, and adding them to the front-panel switching options would increase the complexity very noticeably. Such modes can, however, be called up by the computer without regard to front panel switching. Each mode can be initiated by sending a specific MODE to the OUTPORT:

MODE	OUTPORT
WEFAX	6
APT	5
240 LPM	4
120 LPM	3
NOAA Vis	21
NOAA IR	22
METEOR	23

The manual modes (**APT**, **240 LPM**, and **120 LPM**) are selected from the **MANUAL MODES** section of the **MAIN MENU**, while the **AUTO MODES** section includes **WEFAX**, **NOAA Vis**, **NOAA IR**, and **METEOR**. Mode setting is accomplished in Module 03 (lines 69Ø-80Ø), with line 69Ø used to send the appropriate mode code to the 1700. This section also selects the appropriate string descriptor for the mode in question (lines 70Ø-76Ø), so that when the routine eventually returns to the **MAIN MENU**, the proper mode label is posted.

What happens next is a function of the selected mode. In the case of the manual modes, the computer is expecting the phase function to be activated, so the system will branch to the first part of the phase routine (Module 04; lines 84Ø-102Ø). The scan converter begins to display the incoming image immediately, and the computer monitor displays a menu inviting you to use key 0 to start phasing or **Q** to exit (lines 84Ø-92Ø). If you press **Q** (line 90Ø), the system sends a 255 to put the 1700 into standby and returns you to the **MAIN MENU**. If you press 0 (line 91Ø), you'll be routed to line 93Ø. This line send a 48 to the scan converter, a code that initiates the phasing sequence, just as if you pressed and held the front panel **PHASE** switch. A second menu is posted, inviting you to press 0 to stop the phasing (equivalent to releasing the **PHASE** switch), or **Q** to abort the sequence. As in the previous example, a **Q** (line 100Ø) abort display, while a 0 (line 101Ø) causes a 4 to be sent to the 1700 to stop phasing. At this point, the program returns to the **MAIN MENU** while display commences.

The selected mode is displayed and you can abort display at any time by using the **R** key (<R>eset All Modes). When the computer detects that the image is done, the menu is updated to delete the mode selection, and the scan converter is switched to standby (line 224Ø in the keyboard module).

If the selected mode is one of the automatic polar-orbiter formats (NOAA Vis, NOAA IR, or METEOR, the program branches to line 104Ø, and displays an auto-phase menu, while the scan converter attempts to phase the image. You can exit from this routine, resetting the scan converter by pressing **Q**. Once phasing has been achieved, the system returns to the **MAIN MENU** where the current mode is posted until image display is complete. You can use the **R** key to reset during display, if desired.

If you initially selected **WEFAX**, the system reverts immediately to the Main Menu with a WEFAX-mode notice. The scan converter then displays each new WEFAX frame until you reset the system using the **R** key.

In effect, the program gives you access to all the normal front-panel **MODE** functions, with the added bonus of the NOAA Vis, NOAA IR, and Meteor auto-phase routines.

Saving a 1700 Image

Getting the current 1700 image and saving it to a disk file is accomplished in Module 06 (lines 124Ø-162Ø). Keying an **S** from the **MAIN MENU** (<S>ave Image under the 1700 IMAGE heading) initiates the save routine with the posting of the **Save Current Image** menu and a prompt to insert a file disk. All file disks go in the default drive (the one you used to boot the program). A blank (formatted) 360-kbyte disk will hold five images. Assuming you continue, you'll be prompted for a file name (up to eight characters long). The program automatically appends a .WSH extension to the file name you select.

Whenever the **MODE** switch of the scan converter is in the **HOLD** position, the unit will accept a request from the external computer to send it the current image in the scan converter's memory. This is accomplished by sending a 16 to the OUTPORT as shown in line 144Ø. For reasons that you'll see shortly, we need to keep track of the last video value as the transfer progresses and, here at the start, we also need to zero the memory pointer in our video-data segment. In the case of the previous video value, we only need a dummy at the moment, so the video value (V) is simply set to 0 in line 143Ø.

Now we need to read the output port of the scan converter (your input port) to see if the scan converter has the first pixel ready (line 145Ø). If the scan converter has not yet acknowledged our request, it has a 255 at the output port, the normal standby condition. In such a case, we want to look again (line 146Ø). When the scan converter acknowledges the send request, it does so by echoing back the 16 to your input port. A 16 at your input shows that the scan converter has the send message, but it also doesn't have the first pixel yet—requiring that we go back to 145Ø for another look (line 147Ø).

If the value is *not* 255 or 16, it must be the first data byte. Since we are now ready to actually exchange data, line 148Ø is used to set up a loop for receipt of all 65536 pixels that make up the image.

The rest of the transfer routine is based on the fact that the scan converter sends pixel data such that no two sequential bytes are precisely the same. The reason for this is that the 4-bit data are in the high 4 bits of the byte, while the lower 4 bits are used for handshaking data in the form of either a 1 or a 2, indicating which pixel of a given pair is being sent:

	First Pixel	Second Pixel
Bit 7	V3 (MSB)	V3 (MSB)
Bit 6	V2	V2
Bit 5	V1	V1
Bit 4	V0 (LSB)	V0 (LSB)
Bit 3	low	low
Bit 2	low	low
*Bit 1	low	high
*Bit 0	high	low

Even if the actual video values (V0-V3) of the two pixels are identical, the bytes will always differ because of the different values posted in bits 0 and 1. The entire transfer strategy involves simply comparing the present value with the last video value to determine if the pixel is a new one or not.

The value of the byte is read in line 149Ø and assigned to variable X. In 15ØØ, we compare X with 255 since that is what the scan converter sends if it reaches then end of its RAM before we finish our loop. If a 255 is noted, we terminate the transfer. In line 151Ø we compare the new reading (X) with the previous value (V). If they are the same, this cannot be a new pixel value and we return to 149Ø to look again.

If the values are different (indicating a new byte value), we re-read the value in line 153Ø, assigning it to variable V. This extra step accomplished two tasks: It avoids getting an ambiguous data reading during a transition at the output port of the 1700, and it also automatically assigns the current byte value to the variable we will use when comparing for the next pixel transition. We then store V in RAM (line 154Ø) and echo it to the 1700 (line 155Ø) as a signal that it has been received. We then loop back (line 156Ø) for another comparison in line 149Ø. Meanwhile, the 1700, detecting that the last byte it sent was echoed, fetches the next RAM value and sends it to the computer INPORT.

This transfer continues until either the computer loop count is complete or the 1700 reaches the end of its own RAM storage (usually the latter), at which point the computer sends out a 255 (line 158Ø) to terminate the transfer. The save routine is terminated when the computer saves all 64 kbytes of segment 6000h as a disk file (line 159Ø) under the file name you selected. The program then resets to the default value for the data segment (line 161Ø) and returns to the **MAIN MENU** (line 162Ø).

The principal disadvantage of BASIC in such an image transfer is the time required to handle all 65,536 pixels. On my 10-MHz 80286 PS/2, the routine takes about three minutes using the program shown in Appendix III. The advantage is the relative simplicity and universality of BASIC, because it can be implemented on any computer system. With the version of the program distributed on the *WSH Program Disk*, where the actual image transfer is accomplished in assembly language, the transfer takes 10 seconds on the same computer, including the time required to write the image to the file disk.

Sending an Image to the Scan Converter

Assuming you understand the sequence involved in getting an image from the scan converter, sending an image back from a disk file is just a matter of some role reversal. The transfer is accomplished in Module 07 (lines 166Ø-208Ø). Lines 166Ø-19ØØ are overhead to set up menus, prompt you to insert a file disk, display of all files with a .WSH extension, and your selection of a file to load. When a file has been selected, it is transferred from disk to segment 6000h. The actual routine to move the image data from this RAM segment to the 1700 begins with line 191Ø.

The transfer is initiated by sending a 32 to the 1700 (line 191Ø). When the 1700 is ready to receive the first pixel, it echoes this value. Line 195Ø sets up the loop to fetch all the RAM values. 197Ø is used to get the value from RAM and it is sent to the 1700 in line 198Ø. The program then checks the input port until the 1700 sends back the same byte value—a signal that the scan converter has received and processed the last pixel/byte sent by your computer. Unlike the previous routine, where nothing seems to be happening during the transfer, you'll see the data from the computer being written to the 1700 display as the transfer progresses. As in the previous transfer routine, if the 1700 fills up its video RAM prior to completion of the MEM loop, it outputs a 255, which causes the transfer to abort. When this happens (or the loop is complete) the system resets to the default segment (line 2Ø6Ø), outputs a 255 to verify the end of an image (line 2Ø7Ø) and returns to the **MAIN MENU** (line 2Ø8Ø).

As in the previous example, transfer requires about three minutes on my system using the BASIC version, while only about 10 seconds are required for the machine-language routines used on the *WSH Program Disk*.

General Notes

The syntax and structure of the two previous coding examples are based in PC compatible computers that require segment management and where there is ample RAM available. Computers based on the Motorola 60xxx family feature non-segmented RAM and the DEF SEG statements would not be required. Older 8-bit computers are often limited to 64 kbytes of total RAM, thus presenting a problem because the image files created by the two previous examples require 64 kbytes. Two approaches can be taken. You can strip the lower 4 bits from each incoming byte and pack two pixels per byte, much as the scan converter does, reducing the file size to 32 kbytes. The only disadvantage to this technique is that the two pixels have to be reformatted when sending them back to the scan converter.

Alternatively, the incoming data can be handled in blocks of 16 or 32 kbytes, and more than one disk file can be used to hold a single image. Because image transfer in either direction will not proceed unless you acknowledge each byte, you can suspend the transfer for any length of time to access disk drives or perform other operations. A third alternative is to write each byte to the disk (or read it, in the case of sending an image to the scan converter) as it is received. This requires very little RAM overhead in the host computer, but adds considerably to the time required for image transfer, and also greatly increases the required disk-file size.

Although the programming examples have been presented in BASIC, the same structural logic can be applied to routines written in any language.

High-Resolution Data

The previous discussions have all dealt with transfer of the basic $256 \times 256 \times 4$-bit images as stored by the scan converter. The second major category of image handling is the high-resolution data streamed by the scan converter when it is acquiring an image for display. Irrespective of the actual image format, the streaming data have the image format and data structure outlined below.

Standby

When the scan converter is not loading an image or handling one of the previously described image-transfer routines, it sets all output bits (your computer input) high, resulting in a port value of 255.

Image Start

When the scan converter is about to start loading an image, it sets all output bits low, resulting in a port value of 0. This occurs with the receipt of a valid start tone in the automatic WEFAX mode, when the **PHASE** switch is activated in any of the manual modes (APT, 240 LPM, and 120 LPM, or with the start of autophasing in the case of the automatic modes (NOAA Vis, NOAA IR, and METEOR). This 0 output value can serve to signal the external computer that image streaming is about to begin. After an appropriate WEFAX start delay, after the **PHASE** switch is released in the case of one of the manual display modes, or with completion of autophasing, data streaming begins.

Image Data Stream

The scan converter streams 1024 pixels per line. Pixels are sent in sequential pairs with the following format:

	Data Bits							
	0	1	2	3	4	5	6	7
Pixel #1	H	L	V	V	V	V	V	V
Pixel #2	L	H	V	V	V	V	V	V

The actual video data bits (V) are in a 6 bit format with bit 7 as the MSB and bit 2 as the LSB. Note that the coding in the lower 2 bits (0-1) assures that the byte value for two pixels in a sequence will always be different, irrespective of the actual video value. There is no hand-shaking in this format because the 6809 microprocessor is too busy formatting data and keeping track of time to take the time to determine if your computer is keeping up with the data stream.

Whether or not a computer is connected to the scan converter, it streams 512 pixel pairs (total = 1024) for each line for a total of 768 lines. When all the lines are complete, the scan converter sets all its output bits high (byte = 255) to signal the end of the image. If the operator terminates the image prior to completes (**MODE** switch to **HOLD**), the scan converter posts the 255 (end-of-image) code when the switch is placed in **HOLD** once the current line is completed.

The streaming data are output so rapidly that it's not practical to obtain this data using a slow language such as BASIC. Compiled C code can be made fast enough and assembler is ideal. This data stream can be taken in its entirety or sub-sampled to yield data at any resolution below the $1024 \times 768 \times 6$-bit format provided by the scan converter.

COMMERCIAL SOFTWARE

A compiled version of the WSH1700 program in Appendix III, using fast assembly language routines for image transfers, is contained on the *Weather Satellite Handbook Program Disk* available from the ARRL (see Appendix I). This program is an excellent entry-level option for those of you with PC-compatible computers. It supports all functions, with the exception of the high-resolution streaming, and is extremely easy to use. Since the scan converter is used as the display terminal, the graphics capability of your computer system is irrelevant. The same disk also contains an advanced tracking program and represents a very cost-effective software package.

VGA1700 Program

For IBM compatible systems equipped with VGA display systems, the VGA1700 program from Metsat Products ($50) provides access to the high-resolution data stream provided by the scan converter. The 512×480 display resolution provided by this program was used to obtain most of the satellite images that illustrate this volume.

Because the program is using the 6-bit data streamed by the scan converter, considerable latitude for image processing is available. The VGA1700 program contains five resident image-processing curves (which are discussed in Chapter 10) and supports as many custom curves, stored in external disk files, as you care to use.

The program supports save and loads of the high-resolution images in three disk formats:

- 360 kbytes—2 images/disk
- 720 kbytes—5 images/disk
- 1.44 Mbytes—10 images/disk

In addition to basic image-handling functions, the program also lets you use your computer display as a digital oscilloscope to permit precision level setting.

Although the program is optimized for high-resolution image display, any high-resolution image can be passed back to the scan converter for display at the scan converter's 256 × 256 resolution. This is a useful feature when it is not convenient for larger groups to view the VGA screen—a conventional monitor, connected to the scan converter, can be used for display. Contrast enhancement and other image processing performed on the pictures are displayed on the scan converter, adding to the versatility of this option. In addition to these functions, all of the control functions for manual and automatic modes as well as scan-converter display functions are supported as well. In addition, the program supports routines for unattended image acquisition and disk storage that are outlined in Chapter 10.

Metsat Products is quite cooperative in assisting software developers who want to write scan-converter software for any family of computers. The information provided in this chapter should provide most programmers with the needed information to support their creative efforts, but if you have questions, Metsat will endeavor to assist you. The company will also maintain a listing of all software products that they are aware of that support the scan converter.

SYSTEM EXPANSION

Given the basic scan converter, future expansion is primarily a matter of upgrading your computer capability. There are already VGA cards that support 1024 × 768 × 16 gray-scale display, permitting you to take advantage of the full resolution inherent in the streaming-data format. There is not much point in going further with display resolution because this level represents something very close to the theoretical resolving limit of the satellite direct-broadcast formats (see Chapter 4). The primary advantages of very high-resolution display is the potential to merge

Computer Output Port Value	Model 1700 Scan-Converter Function
	Display:
2	Display
1	Invert
7	Complement
	Mode Control:
6	Automatic WEFAX
5	Manual APT
4	Manual 240 LPM
3	Manual 120 LPM
21	Automatic NOAA visible
22	Automatic NOAA IR
23	Automatic 120-LPM METEOR
	Image Transfer:
16	Send 1700 image to computer
32	Send computer image to 1700
	Miscellaneous:
8	Stream data to computer
48	Start phase routine
255	Abort/standby

Figure 7.4—A summary of the computer-generated control codes recognized by the Metsat Revision 3 ROM. All the functions noted here are discussed in the text with the exception of the Stream data to computer function (under Miscellaneous). This code causes the computer to stream 4096 samples/second to the computer input port for use with oscilloscope-type displays of image levels, or other functions not supported in the standard software routines. Other functions will almost certainly be added to future ROM releases, but when this happens, they'll have new control codes. The codes/functions discussed earlier will be retained to assure backward compatibility with older versions of computer software.

multiple images to produce full-disc pictures from WEFAX data, or mosaics from polar-orbiter data. The latter is a real challenge in terms of image-data management, but well within the capabilities of anyone willing to work hard on the project. A number of avenues for experimentation with your computer/scan converter combination are discussed in Chapter 10.

Chapter 8

Satellite Tracking

INTRODUCTION

In this chapter, you'll be introduced to the basic skills of satellite tracking. There's really nothing complicated about the process—you can do it all with a pencil, paper and elementary school math. Though the math is really quite simple, use of a pocket calculator helps to minimize the chance of error with repetitive calculations. In fact, there is no better subject than satellite orbits to introduce students to the tremendous predictive power of mathematics! Sure, most of the operations we'll cover here can be bypassed by the use of relatively simple computer programs. But if you jump right into tracking satellites by using a computer, you'll miss the chance to really understand the nature of satellite orbits. I'm firmly convinced that everyone can benefit from some experience with manual plotting and tracking techniques, and strongly urge you to acquire the skill—even if you intend to use your computer for day-to-day predictions.

ORBITAL-PLOTTING BOARD

One of the fundamental aids for any satellite station in determining satellite orbital tracks is a plotting board. Figures 8.1-8.3 provide the basic materials you'll need to construct such a board. You'll actually be making *two* plotting boards. One board you'll use in the sample exercises we'll step through; the other will be customized for your own station location. Here's how to create these two versions of the plotting board:

1) Make two photocopies of Figure 8.1. Glue each copy to a piece of poster board, or cardboard, to provide stiffening.

2) Make one copy of Figure 8.2. Use a pair of sharp scissors to carefully cut out the two circles, being careful to retain the N and S labels.

3) If you live in the northern hemisphere, make one copy of Figure 8.3, and use typist's opaquing (correction) fluid (or white paint) to cover up the numbers 51 to 101, adjacent to the points on the track. If you live in the southern hemisphere, make two copies of

Figure 8.3, cover the numbers 51 to 101 on one copy, and 0 to 50 on the second copy.

4) Take your copies of Figure 8.3 and have additional copies made on overhead-transparency film. (Most copy centers have this capability.) If you live in the northern hemisphere, a single copy is adequate. Make transparent-film copies of both of your copies of Figure 8.3, if you live in the southern hemisphere. If you don't have access to copy facilities that handle film material, use india ink or other indelible ink to trace the track arc, points, numbers, and the center cross on a moderately stiff piece of transparent acrylic or translucent Mylar sheet. There is no need to trace the circular or square border areas.

5) Use sharp scissors to cut out your film copy(s) around the circular margin, discarding the borders.

Now let's take a look at the pieces we have created and how they go together to make a plotting board. Figure 8.1 represents a polar projection of the earth, where the center point is either the north or south pole, depending on your location. The radial lines represent lines of longitude. The line coming straight down from the center point is 0°. Note that the longitude radii are labeled in 10° increments from 0 around to 350. Labeled in this manner, the longitude calibrations represent west longitude.

The concentric lines represent lines of latitude, beginning with the equator (0°) as the outermost circle, and working inward in 10° increments to the pole.

The diagram in Figure 8.2 represents azimuth bearings (N, NE, E, SE, S, SW, W, and NW), and elevation circles in 10° increments from 0° at the outermost circle, to 70°, with the center point representing 90°. Attach this diagram to the plotting board at our station location. For our sample tracking exercise, we'll deal with a hypothetical station at 80° W longitude and 40° N longitude. Take one of your copies of Figure 8.1 and use the point of a straight pin to mark this point on the diagram. Coat the back of one of your diagrams of Figure 8.2 with contact cement or other glue and insert the straight pin through the center of the diagram. Insert the point of the pin into the 80° W/40° N

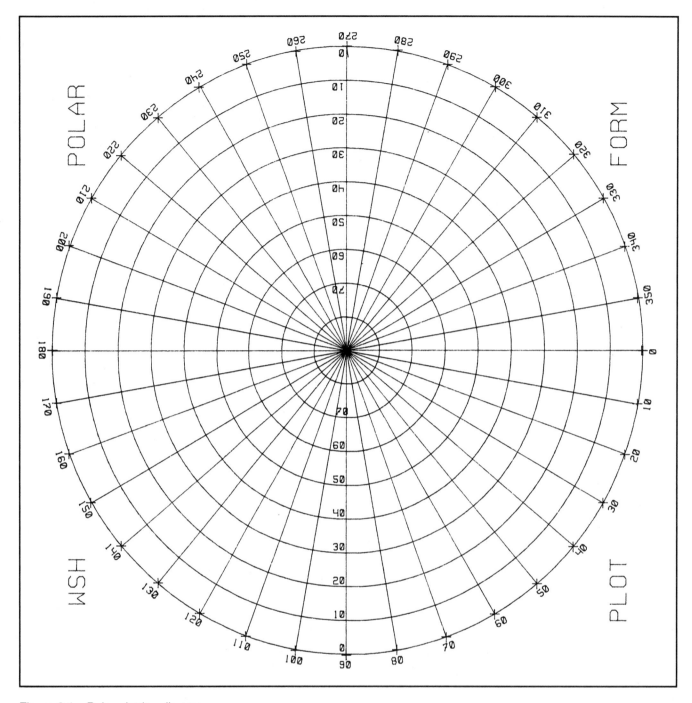

Figure 8.1—Polar plotting diagram.

point you previously marked. Slide the bearing circle down the pin, orienting it so that the N line points directly toward the center of the diagram. Smooth the bearing diagram into place. Your plotting board should now look like Figure 8.4, disregarding the plotting track.

Repeat these steps using your own station location on your second copy of Figure 8.1. If you are located in the southern hemisphere, consider the concentric latitude lines to represent south latitude, and the center point is the south pole. When placing the bearing diagram for a southern-hemisphere station, orient the diagram so that N points toward the pole. When the station diagram overlay has dried, relabel N on the diagram as S and S on the diagram as N.

To avoid smudging and soiling your own plotting board, cover the entire diagram (complete with the bearing circle at your station location) with clear

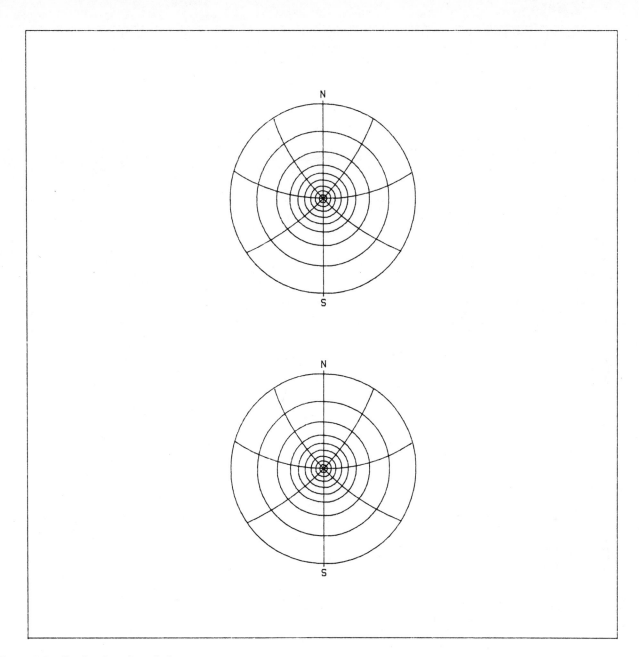

Figure 8.2—Station-bearing circles.

plastic film. There is no need to do this with the sample plot for our station at 80° W/40° N because we'll only use this version for a few sample exercises.

You should have one film copy of the track from Figure 8.3, with the track points labeled 0 to 50. Place a thumbtack through the center-cross of the film track, and insert the tack into the center of the diagram you prepared for our hypothetical 80° W/40° N station. You should now be able to rotate the track diagram around the pole. Set the diagram aside until we're ready for a sample tracking exercise. Once our sample exercises are complete, northern-hemisphere stations will transfer the track film to their own plotting board. Trackers in

the southern hemisphere should attach the second copy of the track diagram (with points labeled 51 to 101) to their southern-hemisphere plotting board.

SATELLITE ORBITS

A satellite orbit is the path traced by the spacecraft in its trip around the earth. There are three major components that define the nature of this orbit: its eccentricity, period, and plane. *Eccentricity* defines the degree of departure of the orbit from circularity. Virtually all natural-object and satellite orbits are elliptical to some degree. In the case of an elliptical orbit, there is a point along the orbital path where the

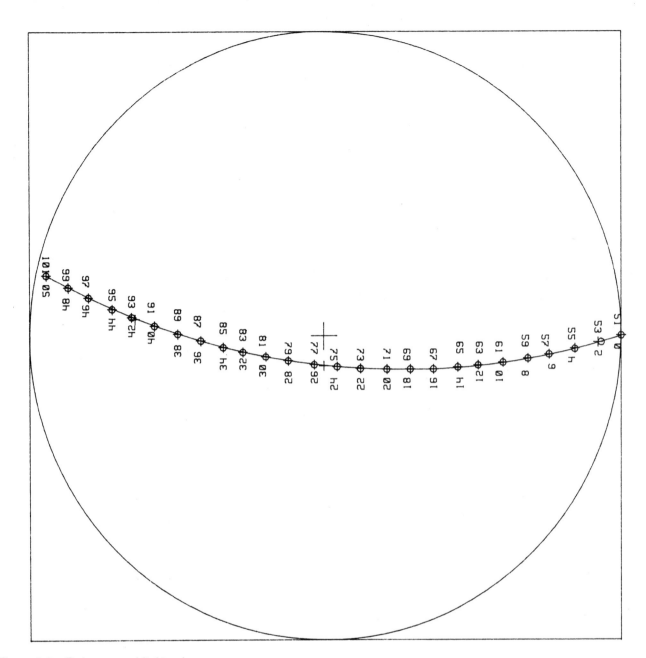

Figure 8.3—Reference orbital track.

satellite is closest to the earth (known as the *perigee*), while 180 degrees around the path is the *apogee*, or the point at which the satellite is farthest from the earth. For an operational weather-satellite satellite system, it is highly desirable to minimize the eccentricity (the ratio of apogee to perigee) and achieve something as close as possible to a circular orbit (apogee = perigee, eccentricity = 1). Although there is no such thing in nature as a perfectly circular orbit, injection of the TIROS/NOAA spacecraft into orbit is controlled as precisely as possible to yield orbits that are circular to within ±18.5 km during the operational lifetime of a given satellite.

The *period* of a satellite orbit is the time, in minutes, required for the satellite to complete one circuit of the earth. The period of the orbit is a function of the satellite's altitude. For near-circular orbits, the relationship can be expressed by:

$$P_{min} = 84.4 \times (1 + (h/r))^{3/2}$$

where P is the period (in minutes), h is the height, or altitude, of the orbit and r is the radius of the earth. As altitude increases, so does the period. The TIROS/NOAA satellites have an average altitude of about 854 km (460 nautical miles [nmi]), resulting in a period of 102 minutes. Geostationary satellites (such as GOES) orbit at an altitude of 35,790 km (19,312 nmi), resulting in a period of 1440 minutes or 24 hours. As a more extreme case, consider the earth, orbiting 93,000,000 miles "above" the sun. The earth follows precisely the same physical laws and requires approximately 365 days to complete an orbit as a satellite of the sun.

The *plane* of the orbit describes the relationship of the plane of the orbital path relative to the earth. An *equatorial* orbit is one in which the plane of the orbit lies in the plane of the equator. (This means that the orbital track lies directly over the equator.) Equatorial orbits are characteristic of the geostationary satellites such as GOES, METEOSAT, and GMS. Recall that these satellites have a period of precisely 24 hours (1440 minutes). Because the direction of their movement is the same as the direction of the earth's rotation beneath them, the satellites remain over the equator at the same point at all times—hence the term *geostationary*.

The mission of weather satellites in lower orbits is to photograph as much of the earth's surface as possible in the course of a day. Equatorial orbits don't achieve this goal at lower altitudes because so much of the earth north and south of the equator is out of view of the satellite. The ideal plane of a weather-satellite orbit is *polar*, where the plane is oriented 90° to the equator, such that the orbital track crosses the poles. As the earth rotates beneath the satellite, virtually all parts of the earth can be photographed during a day with the proper combination of altitude and period.

Precise polar orbits are not desirable if we want the satellite to pass overhead at roughly the same time each day. To achieve this, we require what is known as a *sun-synchronous* orbit. Power and thermal constraints rule out an orbit that produces passes within 2 hours of local noon (or midnight). The result is that TIROS satellite orbits have an inclination of 98.8° to the equator. This results in an orbital track that reaches a latitude of 81.2° (180 – 98.2)—what we refer to as a *near-polar* orbit.

If we're to predict where a satellite will be at a given time, we need some means to graphically present the orbit. At any point in its orbit, there is some point on the earth's surface that is directly beneath the satellite. This point is known as the satellite *subpoint*. If we plot the position of the subpoint at regular intervals during an orbit, we end up with an orbital track projected on to the earth's surface.

To deal with this orbital track in a predictive way, we need a place to start. The standard way to define the "start" of an orbit is to use the point at which the satellite crosses the equator in a north-bound direction. For any given satellite, this equatorial crossing occurs at a specific longitude and time on any given day.

PREDICT-DATA FORMAT

In order to predict where the satellite will be, we need data on a *reference crossing* for the satellite of interest. One of the most successful modes for disseminating this information over the years has been in the form of "Predict" post cards, mailed out each month by the satellite service. These cards (see Chapter 9 for more information) contain data on a single reference crossing (usually for the first orbit on the first day of the month) for each of the operational TIROS/NOAA satellites. An example, the following data was distributed for NOAA-10 for 01 December 1989:

	NOAA-10
ORBIT #	16642
EQ. XING TIME(Z)	0124.96Z
LONG. ASC. NODE(DEG)	88.89W
NODAL PERIOD	101.2340
FREQ. (MHZ)	137.50
INCR. BTWN ORBITS	25.31

ORBIT # represents the number of the reference orbit where orbit numbers are tabulated from the first orbit at launch.

EQ. XING TIME(Z) is the time when the crossing of the equator occurs. The NOAA postcard format treats the time format as follows:

HHMM.MM

where H is hours and M represents minutes to two decimal places. As you can see, orbit 16642 begins at 01 hours and 24.96 minutes.

As in all aspects of satellite work, the time is referenced to that at the prime meridian (0° longitude). The Z stands for Zulu time, a commonly used designator. Greenwich Mean Time (GMT) is another common (though obsolete) designation for prime-meridian time. The proper term to use is UTC. (Because of governmental use of Z, that letter and UTC will be used interchangeably here.) Although you can convert UTC to your local time, such conversions simply serve to increase the chance for error in your orbital predictions. Virtually all satellite stations maintain at least one clock set to UTC. The universal availability of inexpensive digital clocks makes this quite convenient, and provides a degree of accuracy more than sufficient for our purposes. You'll soon be adept at making conversions in your head in so far as knowing when the satellite's timetable interacts with your own personal schedule.

If you want to convert the decimal minutes value to seconds, simply multiply the decimal component (0.96 in the case of our NOAA reference data) by 60. Thus, in terms of hours, minutes, and seconds, our reference orbit (16642) begins at 01:24:58Z.

LONG. ASC NODE (DEG) refers to the longitude of the ascending node (in degrees); a way of stating the point (in degrees West longitude) where the satellite crosses the equator northbound at the start of orbit 16642 at 01:24:58Z—in this case, 88.86° W.

PERIOD represents the time (in minutes, to four decimal places) required for the satellite to complete each orbit. In the case of NOAA-10, each new orbit begins 101.234 minutes after the start of the last orbit. There are many factors that can effect the value of the period. Because the TIROS/NOAA satellites are at relatively low altitudes, residual atmospheric drag is a factor (which varies under the influence of many factors, including solar activity) that tend, over time, to slow the satellites, resulting in a slightly lower orbit and lowering the period. These variables are the primary limitation on how accurate we can be in simple, long-term predictions—a factor I'll discuss at greater length later in this chapter.

The FREQ. entry is simple: It tells us what transmitting frequency the satellite is using.

INCR. BTWN ORBITS is a bit obscure. It refers to the longitude increment between successive orbits. If the earth were not rotating, by the time the satellite had completed one orbit and was ready to start another, the new crossing would start at precisely the same subpoint on the equator as did the previous orbit. But, the earth *is* rotating at a rate of 360° every 24 hours (1440 minutes), or 0.25° per minute to the east. In the course of the 101.234-minute period required for one NOAA-10 orbit, the earth has rotated beneath the satellite a total of 101.234 × 0.25, or 25.3085 degrees toward the east. Because, within the course of a day, the satellite orbital plane is fixed, the next crossing point will be 25.3085 degrees to the *west* of the previous crossing. It is this value, rounded to 25.31, that represents the longitudinal increment between successive crossings. Because long-term predictions involve adding the increment repeatedly with each orbit, it is important that the increment be as accurate as possible. For this reason, I prefer to use the more accurate value derived from the period (as shown above) in preference to the rounded value included on the predict card.

With the combination of data provided on the Predict card and the concepts presented in the previous paragraphs, we have all the tools in hand to track the satellite in time—the subject of the next section.

PREDICTING CROSSINGS

Given reference crossing data of the type we have for NOAA-10, predicting each subsequent crossing is a simple matter:

- Add the period to the previous crossing time to get the new crossing time.

- Add the increment between orbits to the old crossing point to get the new crossing point.
- Increment the orbit number.

Before we actually go ahead and do this, there is one useful shortcut we can take that makes the job easier. Although we can add hours, minutes, and seconds for the time, the process is error-prone because calculators are not designed for the job. Things are much easier if we convert our reference crossing time to minutes:

$$T_{min} = (H \times 60) + M$$

Because our reference time for orbit 16642 is 01 hours and 24.96 minutes:

$$T = (1 \times 60) + 24.96$$
$$= 84.96 \text{ min}$$

Now we can add the period for each new orbit, keeping the time in the form of total minutes. We'll only bother to convert the time back to the HH:MM:SS format when needed—something we won't have to do with every orbit, as you'll see.

Next, take a piece of lined paper and make three columns, labeled ORBIT, TIME, and CROSSING. In the ORBIT column, enter 16642 to 16656 in sequence. The reason for stopping at 16656 will be evident shortly.

Now, load the value for period (101.2340) into your calculator's memory. This is usually done by keying in the number and pressing the M+ key on most calculators. Next, key in our starting time (84.96), and keep adding the value in memory (using the RM key) to the running total, recording the value in each succeeding TIME column until you reach ORBIT 16656. Your tally sheet should look something like this at the moment:

Orbit	Time (min.)	Crossing (°W)
16642	0084.960	
16643	0186.194	
16644	0287.428	
16645	0388.662	
16646	0489.896	
16647	0591.130	
16648	0692.364	
16649	0793.598	
16650	0894.832	
16651	0996.066	
16652	1097.300	
16653	1198.534	
16654	1299.768	
16655	1401.002	
16656	1502.236 (0062.236)	

Note that for orbit 16656, the TIME has reached a total of 1502.236 minutes. Because there are only 1440 minutes in a day, this orbit must occur during the next day—December 2. We can correct this time by subtracting 1440, yielding 62.236 for the first orbit on December 2. Assuming our starting orbit of 16642 is the first orbit of the day (it is, because if we subtract the period from the time, we would have a negative number), orbits 16642 through 16655 represent all possible crossings for the 1st of December and we already have the crossing times calculated.

Now we must calculate the crossing point for each orbit. Load the value for the increment between orbits (25.3085) into the calculator memory, key in the crossing point for orbit 16642 (88.86), and start a running total recording the values in the CROSSING column (rounded to the nearest whole degree) until you reach orbit 16653. Note that the total on your calculator is 367.2535. Because there are only 360 degrees in a circle, we have obviously worked our way *around the earth* in our subsequent crossings, until we are just past the prime meridian.

To correct for this, subtract 360 (yielding a value of 7.2535), and continue your calculations and recording of data. By the time you have reached orbit 16656, your sheet should look like this:

NOAA-10 01 December 1989

Orbit	Time (min)	Crossing (°W)
16642	0084.960	089
16643	0186.194	114
16644	0287.428	139
16645	0388.662	165
16646	0489.896	190
16647	0591.130	215
16648	0692.364	241
16649	0793.598	266
16650	0894.832	291
16651	0996.066	317
16652	1097.300	342
16653	1198.534	007
16654	1299.768	033
16655	1401.002	058
16656	0062.236	83

We now have the crossing times and crossing point for all of the orbits on December 1, as well as the first orbit (16656) on December 2. In the next section, we'll look at how to use our plotting board to convert these numbers into real tracking data!

WHERE IS THE SATELLITE?

Now that we know when and where the satellite will make equatorial crossings, we need the plotting board to see what these numbers mean in terms of the satellite track. If you spend much time listening to a specific satellite frequency, such as the 137.50-MHz frequency of NOAA-10, you already know that most of the time you'll not hear the satellite's signal. The reason for this will become clearly evident.

Take the plotting board you prepared for our sample 80° W/40° N station location. Let's look at the satellite track for the first orbit (16642) on our table. Rotate the overlay so the point label 0 is located at 89° W on the polar diagram. Your diagram should look like the one in Figure 8.4.

What your diagram is showing you is the path of the satellite for the first half of orbit 16642, with the subpoint noted every two minutes. The outermost bearing circle around our station location represents 0° elevation—our radio horizon. Unless the satellite track intersects this outer circle, we have no chance of hearing the satellite. Note that the track does intersect the outermost bearing circle, so we'll hear the satellite during this pass.

Because we might actually want to listen on this pass, the first step is to convert the crossing time in minutes back to an hour/minute/seconds format. This is accomplished in three steps:

1) HOURs are equal to the whole number obtained when we divide the total minutes by 60:

$$HOURS = INT(84.96/60) = INT(1.416) = 01$$

2) MINUTES are calculated by multiplying the remainder in Step 1 (0.416) by 60:

$$MINUTES = INT(0.416 \times 60) = 24.96 = 24$$

3) SECONDS are calculated by multiplying the remainder in Step 2 (0.96) by 60:

$$SECONDS = 0.96 \times 60 = 57.6 = 58$$

Thus, by our UTC clock, the crossing can be expected at 01:24:58Z. If we were using the Zapper omnidirectional antenna, we would expect acquisition of signal (AOS) from the satellite at about 5 minutes after the crossing (AOS = 01:29:58). The satellite will pass off to the west of our station (reaching a maximum elevation of slightly more than 15°, and we should get loss of signal (LOS) at about 17 minutes after the crossing (LOS = 01:41:58). This is not a particularly good pass, but if we wanted to record it, we could simply set a digital appliance timer to turn the recorder on at AOS (01:29) and off at LOS (01:42), and the pass would be recorded on tape for later playback.

Note that during orbit 16642, the satellite will approach from the south and disappear to the north. Such a track is known as an *ascending* pass.

If we were using a gain antenna and wanted to track the satellite, our plotting board could provide the

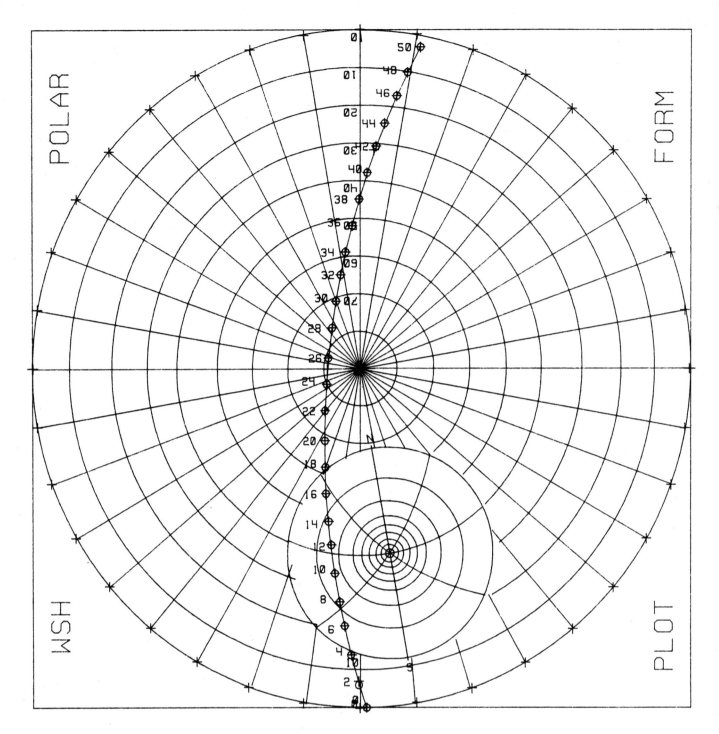

Figure 8.4—Sample plotting board layout for a reference crossing at 88° W.

basic information needed with a few more steps. Prepare a lined sheet with four labeled columns—MAC, TIME, ELEVATION, and AZIMUTH. In the MAC column (minutes after crossing), enter values from 5 to 17 at one-minute intervals. In the TIME column, compute the time equivalent for MAC = 5 by adding 5 minutes to the equatorial crossing (01:24:58 + 5 min = 01:29:58. Fill in the remaining columns through MAC = 17 by adding one minute for each entry. The TIME for MAC = 17 should equal 01:41:58.

Now, using the plotted track (interpolating between the two-minute points), determine the elevation and azimuth at each one-minute interval. The concentric circles in Figure 8.4 represent elevation in 10° intervals from 0° to +70°. The radial lines represent azimuth (direction). When complete, your table should look something like this:

NOAA-10 Orbit 16642

MAC	Time	Elevation	Azimuth
5	01:29:58	4	SW
6	01:30:58	6	SW
7	01:31:58	9	SW
8	01:32:58	11	WSW
9	01:33:58	15	WSW
10	01:34:58	16	W
11	01:35:58	17	W
12	01:36:58	16	WNW
13	01:37:58	13	WNW
14	01:38:58	10	NW
15	01:39:58	7	NW
16	01:40:58	5	NW
17	01:41:58	2	NNW

If you were to set the antenna to the indicated elevation and azimuth values at each one-minute interval (either manually or with an az-el rotator system), you would successfully track the satellite during the time it was above the horizon.

A note here is appropriate for real purists. Our bearing diagram that is placed at the station location on our plotting board is simplified to a considerable degree. Although the elevation circles are plotted as circles, they have shapes that vary with station latitude. Similarly, the azimuth bearing lines are shallow arcs whose shape varies with station latitude. In the case of our bearing circle, the azimuth arcs are correct for 40° latitude. If we were using a very-high-gain/narrow-beamwidth antenna, the errors resulting from such simplifications could be significant. In the case of small 4- and 5-element beams, however, our simple, general-purpose bearing diagram is more than accurate enough to maintain a noise-free signal. If extreme tracking accuracy were required, in moving a small dish to acquire the HRPT signal for example, it would actually be easier to use one of the more elaborate tracking programs of the type I'll discuss later in this chapter.

Now, refer to your worksheet showing the other crossings for December 1, and we can proceed to check out the other orbits for the day. Rotate the overlay so the 0 point on the track is aligned with our next crossing point (114° for orbit 16643). Note that this pass is further west of our hypothetical station plotting and that it doesn't intersect the outermost elevation circle at all. We wouldn't hear the satellite at all during this pass at any point. If you proceed with orbits 16644 through 16647, you'll see that all are out of range of our station and thus of no interest. It should be obvious now why we didn't bother with converting all our total minutes times. The only ones we need bother with are the ones associated with passes that bring the satellite above our local horizon.

If you check out orbit 16648 (241°), you'll see that it intersects the outermost elevation circle to the east of the station, but the maximum elevation is very low. Skip on to orbit 16649, and your plotting board should look like that of Figure 8.5. Make up a tracking worksheet like you did for orbit 16642, and attempt to calculate the times and antenna bearings as you did for 16642. Note, however, some crucial differences as you proceed. First, this is a *descending* pass—it originates to the north and you'll lose the satellite as it moves to the south, as opposed to the *ascending* mode illustrated by 16642. Also, pay close attention to the MAC times on the track plot. In this case, AOS occurs about 31 minutes after the reference crossing, with LOS at about 46 minutes after the crossing. If you have the idea, your worksheet should look something like this:

NOAA-10 Orbit 16649

MAC	Time	Elevation	Azimuth
31	13:44:36	0	N
32	13:45:36	4	N
33	13:46:36	8	N
34	13:47:36	11	N
35	13:48:36	17	N
36	13:49:36	24	NNW
37	13:50:36	30	NNW
38	13:51:36	40	NW
39	13:52:36	42	WNW
40	13:53:36	36	W
41	13:54:36	28	WSW
42	13:55:36	20	WSW
43	13:56:36	15	SW
44	13:57:36	9	SW
45	13:58:36	4	SW
46	13:59:36	0	SW

Note that this is the best pass we have seen so far (it is, in fact, the best of the day) because we have the satellite within range for about 15 minutes and it will reach a maximum elevation of about 42°.

After checking out the remaining passes for December 1, you are ready to put aside your sample plotting board and use the one that is set up for your station location. Repeat our tracking exercises for your station location and you are well on your way toward proficiency with the plotting board! Don't discard the sample board yet, however, for there is one more exercise we'll be doing with it after a few more skills have been added to your repertoire.

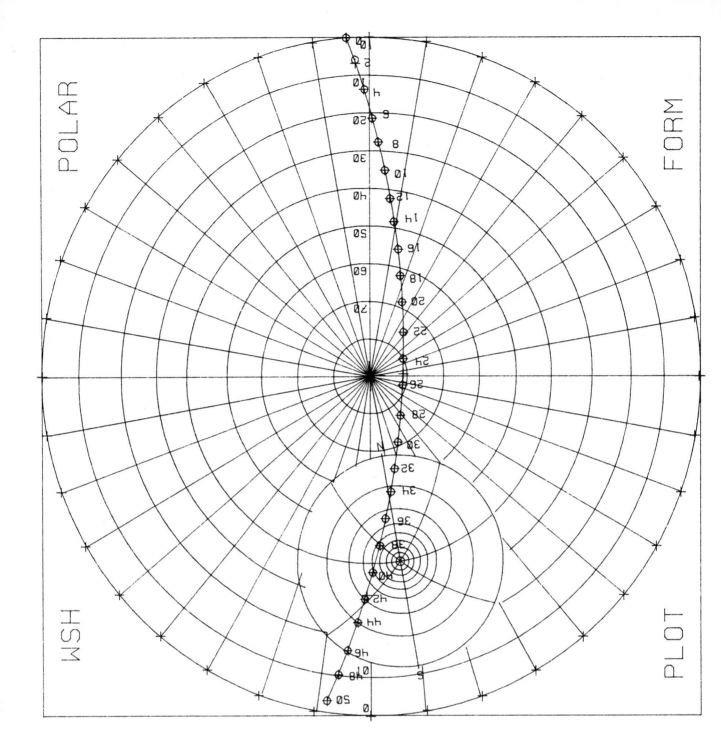

Figure 8.5—Sample plotting-board layout for a reference crossing at 266° W.

SOUTHERN-HEMISPHERE SATELLITE TRACKING

All our exercises so far have been appropriate for northern-hemisphere stations where we are concerned with the *first half* of each orbit. Southern-hemisphere stations, by contrast, are interested in the *second half* of the orbit (minutes 51 to 102). If you followed my earlier instructions on plotting-board preparation, your southern-hemisphere board will have a track overlay with subpoints labeled 51 to 101. Your tracking exercises involve one additional step.

Begin by setting the point on the overlay marked 51 at the crossing point for the orbit in question. This defines the ground track for the satellite during the first half (northern hemisphere) of the orbit. Note the point where the track intersects the equator at the far end of the track (just beyond 101). Mark this point, and rotate the overlay until point 51 is at that point on the equator. Now we have the track in place, properly labeled, for the second half (southern hemisphere) of the orbit and you can evaluate the pass relative to your location.

LONGER-TERM PREDICTIONS

If we were going to move through the month a day at a time, we could simply repeat what we have done previously. That's because the computation of each day's passes automatically yields the first orbit of the next day, which we can use as a starting point. If we had predict data for December 1, and we needed to track the satellite for the 15th, we could go through all the intervening days, but it would waste quite a bit of time. What we really need is a shortcut for jumping ahead by 14 days (15 – 1). The shortcut involves a series of 9 calculations:

1) First, determine the total number of minutes represented by the 14-day differential:

$$1440 \times 14 = 20160 \text{ min}$$

2) Divide the result obtained in Step 1 by the orbital period to determine how many orbits occur in that interval:

$$20160/101.2340 = 199.1426$$

3) To be sure that we'll land in the target day, we will use the next-highest whole number for total orbits: 200.

4) Determine the total number of minutes in 200 orbits:

$$200 \times 101.234 = 20246.8$$

5) Subtract the total obtained in Step 1 from the value in Step 4 to determine how much time must be added to our reference orbit:

$$20246.8 - 20160 = 86.8 \text{ min}$$

6) Add the result obtained in Step 5 to our reference time for orbit 16642 to get the orbital crossing time:

$$84.96 + 86.8 = 171.76 \text{ (new reference time)}$$

7) Add the result obtained in Step 3 to get the new reference-orbit number:

$$16642 + 200 = 16842$$

8) We must convert the time differential in Step 5 to a crossing point differential by multiplying by 0.25 (earth rotation/minute):

$$86.8 \times 0.25 = 21.7 \text{ degrees}$$

9) Add the result obtained in Step 8 to the original crossing point to get the new reference crossing point:

$$88.86 + 21.7 = 110.56 \text{ degrees}$$

In just 9 steps, as opposed to the dreary calculation of 200 intervening orbits, we have a reference crossing for December 15:

Orbit	Time	Crossing
16842	171.76Z	110.56W

Note that because the time is 171.76, while the period equals 101.234 minutes, there is actually one earlier orbit on this particular day. At most locations this may be irrelevant, but you can calculate the values for this earlier orbit by decrementing the orbit number by 1, subtracting the period from the crossing time, and subtracting the interval between orbits (25.3085) from the crossing point. At this point, the remaining orbits for December 15 can be handled just as we did for December 1.

THE CONCEPT OF PASS WINDOWS

It should be obvious at this point that only a small number of orbits each day yield useful passes at any specific station location. It turns out to be quite easy to predict which passes these will be, based on the equatorial crossing point. The best possible passes are the ones where the satellite passes directly overhead. For either an ascending or descending pass, this can only happen if the crossing occurs as a specific point.

Using the sample plotting board, rotate the overlay until the satellite track passes directly over the center point for the sample station location in the ascending mode. If you now examine the location of the 0 subpoint, it should be located at about 69° W longitude. For this hypothetical station location, an ascending overhead pass can *only* occur with a crossing at 69°. While the overlay is in the overhead position, note the times when the satellite will be overhead. AOS will occur at about four minutes after the crossing, the satellite will be overhead approximately 11 minutes after the crossing, and LOS will occur at about 19 minutes.

Although overhead passes are relatively rare, each day you'll have one pass that comes closest to the overhead orientation. Because the increment between orbits is 25.3085 degrees (a value we'll round to 26° for convenience), this "best ascending pass" must fall within ±13 degrees of the nominal 69° overhead crossing point. To be a bit conservative, we could widen this window to ±15°. If we are interested in the best possible ascending pass of the day for *any* satellite in the TIROS/NOAA series, only passes within the window of 54° (69 – 15) and 84° (69 + 15) need concern us. At least one pass must fall within this window during the course of a day. In the case of our December 1 data for NOAA-10, orbit 16655 is the best of the lot for an ascending pass. As far as times are concerned, an AOS of +4 minutes, closest approach to overhead at +11 minutes, and LOS at +19 minutes, derived from the overhead track, are essentially accurate enough for any pass within the best-ascending-pass window.

A similar exercise can be performed for the descending configuration by orienting the overlay to produce a descending overhead pass. This should occur with a crossing of 257°. In this case, AOS occurs at about +32 minutes, overhead at about +39 and LOS at +47. Our best descending window must thus fall between 242° (257 − 15) and 272° (257 + 15).

Summarizing this information for our hypothetical station, we get a short table that looks something like this:

	ASCENDING	DESCENDING
Min. Crossing	54	242
Overhead Crossing	69	257
Max. Crossing	84	272
AOS time	+4	+32
Overhead time	+11	+39
LOS time	+19	+47

If you use an omnidirectional antenna and a timer connected to the station tape recorder, there's no need to consult the plotting board at all to determine which passes to record. To select the best ascending pass of the day, pick the best pass (typically there's only one) that falls in the 54- to 84-degree window, and simply set the timer to go on four minutes before the crossing and turn off at 19 minutes after the crossing. For descending passes, simply scan the calculations for the pass within the 242- to 272-degree window, set the timer for ON at +32 minutes and OFF at +47 minutes, relative to the crossing time.

If you've followed the discussion this far, you can proceed to construct a similar table for your own station location. Both your table and the sample given previously are needed in the discussion of the computer program that follows in the next section.

A SIMPLE PREDICT PROGRAM

It should be obvious that the major problem with manual tracking is the inordinate number of calculations that have to be made, even with a memory calculator. Much of this burden is eliminated by the use of a simple computer tracking program such as the one in Figure 8.6. The program is written in GW BASIC and runs without modification on IBM and compatible PCs. Very little, if any, alteration of the program should be required to run under other BASIC dialects.

Entering the Program Listing

Accuracy is the key to successfully keying in the program. If you are generally unfamiliar with BASIC programming, be aware that small typos can generate large numbers of SYNTAX errors when you first attempt to run the program, and each must be fixed before the program will operate properly. Numerical entry errors may not generate error messages, but can lead to incorrect results. Take your time! As one time-saver, all the REM lines (REMarks) can be omitted on your copy of the program provided you retain the listing for reference as you try to understand what the program is doing. When you have the program entered, save a copy to disk or tape. You'll probably have some editing to do to correct errors, but the saved copy eliminates the need to re-do the job if a catastrophe occurs. As you gradually edit out the inevitable errors, keep saving the program until you finally have a clean master copy.

Language and Computer Compatibility

The program runs on IBM and other PC compatibles with no problem and should work well on most other systems with fairly advanced BASIC features. The only two problems you may run into on other systems concerns variable names and screen formats.

Advanced versions of BASIC allow long variable names (such as PASSWINDOW), and I have used such long names to make the variables as obvious as possible. Some 8-bit home-computer BASIC dialects support only two-character-long variables. If this is the case with your system, go through and assign rational abbreviations to the various variable names (such as PW for PASSWINDOW), being sure to catch all occurrences of each variable and array name. If you know in advance that you'll have this problem, make the changes as you enter the listing, keeping a running list of your abbreviations for reference as you move through the program.

The program is set up for an 80-character-wide screen display. If your computer cannot handle such a display, you'll have to rework the various PRINT statements to fit your available screen size.

There are no unusual functions used in this program, so other programming-language problems are not likely to arise.

Running the Program

The program, as listed, is customized for our hypothetical 80° W/40° N station location, and includes satellite reference data for December 1, 1989. If you first test the program with this data, you can cross-check the results against your manual calculations to verify that all is working well. We'll cover the business

PREDICT.BAS Program Listing

```
1Ø    CLS
2Ø  REM ***********************************************************
3Ø  REM STATION DATA
4Ø  PASSWINDOW = 15
5Ø  ASCLONG = 69
6Ø  ASCOVERTIME = 11
7Ø  DESLONG = 257
8Ø  DESOVERTIME = 4Ø
9Ø  REM ***********************************************************
1ØØ REM READ SPACECRAFT DATA ON FILE
11Ø READ NUMENTRIES
12Ø READ DAY$
13Ø FOR N = 1 TO NUMENTRIES
14Ø READ SPACECRAFT$, ORBIT, TIME, CROSSING, PERIOD, FREQUENCY$
15Ø SPACECRAFT$(N)=SPACECRAFT$:ORBIT(N)=ORBIT:TIME(N)=TIME
16Ø CROSSING(N)=CROSSING:PERIOD(N)=PERIOD:FREQUENCY$(N)=FREQUENCY$
17Ø NEXT N
18Ø REM ***********************************************************
19Ø REM DISPLAY DATA ON FILE
2ØØ REM PRINT BANNER
21Ø GOSUB 159Ø
22Ø PRINT:PRINT
23Ø X$="FILE REFERENCE DATE: "+DAY$:GOSUB 171Ø
24Ø PRINT
25Ø FOR N = 1 TO NUMENTRIES
26Ø N$ = STR$(N):N$ = RIGHT$(N$,1)
27Ø X$ = "("+N$+")  "+SPACECRAFT$(N):GOSUB 171Ø
28Ø NEXT N
29Ø PRINT:PRINT
3ØØ X$ = "KEY NUMBER FOR DESIRED SPACECRAFT OR Ø TO QUIT....":GOSUB 171Ø
31Ø Q$ = INKEY$:IF Q$="" THEN GOTO 31Ø
32Ø CLS:IF Q$ = "Ø" THEN END
33Ø N = VAL(Q$):IF (N>NUMENTRIES)OR(N=Ø) THEN GOTO 19Ø
34Ø REM ***********************************************************
35Ø REM SET REFERENCE DATA VALUES FOR TARGET SPACECRAFT
36Ø SPACECRAFT$=SPACECRAFT$(N)
37Ø ORBIT = ORBIT(N)
38Ø TIME = TIME(N)
39Ø CROSSING = CROSSING(N)
4ØØ PERIOD = PERIOD(N)
41Ø FREQUENCY$ = FREQUENCY$(N)
42Ø REM ***********************************************************
43Ø REM CONVERT NOOA TIME FORMAT TO TOTAL MINUTES
44Ø HOUR = INT(TIME/1ØØ)
45Ø MINUTES = TIME - (HOUR*1ØØ)
46Ø TIME = (HOUR * 6Ø) + MINUTES
47Ø DAY = 1
48Ø REM ***********************************************************
49Ø REM GET TARGET DAY
5ØØ CLS
51Ø GOSUB 159Ø:PRINT:PRINT:PRINT
52Ø X$="REFERENCE DATE: "+DAY$:GOSUB 171Ø
53Ø PRINT:PRINT
```

Figure 8.6—Listing for the WSH PREDICT.BAS program.

```
540  PRINT"DESIRED DAY FOR ";SPACECRAFT$;:INPUT TARGETDAY:CLS
550  REM ***************************************************
560  GOSUB 1590:PRINT:PRINT:PRINT
570  X$="COMPUTING CROSSING DATA":GOSUB 1710
580  REM CHECK IF TARGET DAY MATCHES CURRENT DAY
590  IF DAY = TARGETDAY THEN GOTO 620
600  GOSUB 1030:GOTO 590
610  REM ***************************************************
620  REM PRINT CROSSING DATA
630  CLS
640  REM PRINT BANNER
650  GOSUB 1590:PRINT
660  PRINT "SPACECRAFT:    ";SPACECRAFT$;TAB(50)"FREQUENCY: ";FREQUENCY$:PRINT
670  PRINT "REFERENCE DAY: ";DAY$;TAB(50)"CURRENT DAY:";TARGETDAY:PRINT
680  PRINT"---------------------------------------------------------------"
690  PRINT TAB(10);"CROSSING";TAB(20);"CROSSING";TAB(30);"AOS";
700  PRINT TAB(40);"LOS"
710  PRINT "ORBIT";TAB(10)"TIME (Z)";TAB(20);"DEG W";TAB(30);"TIME";
720  PRINT TAB(40);"TIME";TAB(50);"TRACK"
730  PRINT "---------------------------------------------------------------"
740  REM ***************************************************
750  REM TEST CROSSING AGAINST WINDOW AND PRINT IF FIT
760  GOSUB 1150
770  REM ***************************************************
780  REM COMPUTE NEXT ORBIT
790  GOSUB 1030
800  REM CHECK FOR NEXT DAY
810  IF DAY > TARGETDAY THEN GOTO 830
820  GOTO 750
830  PRINT "---------------------------------------------------------------"
840  X$="NEXT DAY (Y/N)?":GOSUB 1710
850  Q$ = INKEY$:IF Q$ = "" THEN GOTO 850
860  IF (Q$="y")OR(Q$="Y") THEN TARGETDAY = DAY:GOTO 620
870  IF (Q$="n")OR(Q$="N") THEN CLS:GOTO 190
880  GOTO 850
890  REM ***************************************************
900  REM CONVERT TIME IN MINUTES TO H:M:S STRING
910  HOUR = INT(TIME/60):X$=STR$(HOUR):GOSUB 980:HOUR$=X$
920  MINUTES = INT(TIME-(HOUR*60))
930  X$ = STR$(MINUTES):GOSUB 980:MINUTE$=X$
940  SECONDS = TIME -((HOUR*60)+MINUTES):SECONDS = SECONDS * 60
950  SECONDS = INT(SECONDS + .5):X$ = STR$(SECONDS):GOSUB 980:SECONDS$ = X$
960  CLOCK$ = HOUR$+":"+MINUTE$+":"+SECONDS$:RETURN
970  REM ***************************************************
980  REM FORMAT STRING COMPONENTS
990  IF LEN(X$)=3 THEN X$ = RIGHT$(X$,2):GOTO 1010
1000 X$=RIGHT$(X$,1):X$ = "0"+X$
1010 RETURN
1020 REM ***************************************************
1030 REM NEXT ORBIT ROUTINE
1040 REM UPDATE TIME
1050 TIME = TIME + PERIOD:IF TIME<1440 THEN GOTO 1090
1060 REM UPDATE DAY AND INCREMENT PRECESSION
1070 TIME = TIME - 1440:DAY = DAY +1:CROSSING = CROSSING +(360/365)
1080 REM UPDATE CROSSING POINT
```

Figure 8.6—Listing for the WSH PREDICT.BAS program (continued).

```
1090 CROSSING = CROSSING + ((PERIOD/1440)*360)
1100 IF CROSSING >= 360 THEN CROSSING = CROSSING - 360
1110 REM UPDATE ORBIT NUMBER
1120 ORBIT = ORBIT + 1
1130 RETURN
1140 REM ********************************************************
1150 REM WINDOW TEST
1160 C = INT(CROSSING + .5)
1170 REM SET WINDOW BOUNDARIES
1180 AMAX = ASCLONG + PASSWINDOW
1190 AMIN = ASCLONG - PASSWINDOW
1200 DMAX = DESLONG + PASSWINDOW
1210 DMIN = DESLONG - PASSWINDOW
1220 REM CHECK FOR ASCENDING WINDOW FIT
1230 IF (C<AMAX)AND(C>AMIN) THEN GOTO 1270
1240 REM CHECK FOR DESCENDING WINDOW FIT
1250 IF (C<DMAX)AND(C>DMIN) THEN GOTO 1310
1260 GOTO 1580
1270 REM ASCENDING WINDOW
1280 CENTER = ASCLONG
1290 TRACK1$ = "ASCENDING "
1300 GOTO 1350
1310 REM DESCENDING WINDOW
1320 CENTER = DESLONG
1330 TRACK1$ = "DESCENDING "
1340 GOTO 1350
1350 REM DETERMINE TRACK LABEL
1360 IF C = CENTER THEN TRACK2$ = "OVERHEAD":GOTO 1430
1370 IF C>CENTER THEN TRACK2$ = "WEST":GOTO 1410
1380 TRACK2$ = "EAST"
1390 IF C>=CENTER -5 THEN TRACK2$ = "NEAR OVERHEAD "+TRACK2$
1400 GOTO 1430
1410 IF C<=CENTER + 5 THEN TRACK2$ = "NEAR OVERHEAD "+TRACK2$
1420 GOTO 1430
1430 TRACK$ = TRACK1$ + TRACK2$
1440 REM COMPUTE CROSSING TIME STRING
1450 GOSUB 900: CTIME$ = CLOCK$
1460 REM COMPUTE AOS TIME STRING
1470 T = TIME
1480 IF TRACK1$ = "ASCENDING " THEN TIME = T + ASCOVERTIME:GOTO 1500
1490 TIME = T + DESOVERTIME
1500 OT = TIME
1510 TIME = OT-7:GOSUB 900:AOSTIME$ = LEFT$(CLOCK$,5)
1520 REM COMPUTE LOS TIME STRING
1530 TIME = OT+7:GOSUB 900:LOSTIME$ = LEFT$(CLOCK$,5)
1540 REM PRINT ORBITAL VALUES
1550 TIME = T
1560 PRINT ORBIT;TAB(10);CTIME$;TAB(20);C;TAB(30);AOSTIME$;
1570 PRINT TAB(40);LOSTIME$;TAB(50);TRACK$
1580 RETURN
1590 REM ********************************************************
1600 REM PRINT BANNER
1610 X$="*****************************":GOSUB 1710
1620 X$="* POLAR ORBIT PREDICT PROGRAM *":GOSUB 1710
1630 X$="*                             *":GOSUB 1710
```

Figure 8.6—Listing for the WSH PREDICT.BAS program (continued).

```
1640 X$="*              VER. WSH4              *":GOSUB 1710
1650 X$="*                                    *":GOSUB 1710
1660 X$="*        DR. RALPH E. TAGGART        *":GOSUB 1710
1670 X$="*************************************":GOSUB 1710
1680 PRINT
1690 RETURN
1700 REM ************************************************************
1710 REM FORMAT TO CENTER SCREEN DATA (80 COLUMN)
1720 L = LEN(X$):L=INT(L/2)
1730 LEADBLANKS = 40-L
1740 PRINT TAB(LEADBLANKS);X$
1750 RETURN
1800 REM ************************************************************
1810 REM SPACECRAFT FILE DATA
1820 REM NUMBER OF SPACECRAFT ON FILE
1830 DATA 3
1840 REM REFERENCE DATE
1850 DATA "01 DEC 1989"
1860 REM DATA FORMAT
1870 REM SPACECRAFT, ORBIT, TIME(Z), CROSSING(W), PERIOD, FREQUENCY
1880 DATA "NOAA  9", 25601, 130.75, 127.14, 102.0213, "137.62 MHZ"
1890 DATA "NOAA 10", 16642, 124.96, 88.86, 101.234, "137.50 MHZ"
1900 DATA "NOAA 11",  6101, 55.53, 165.32, 102.0851, "137.62 MHZ"
2000 REM ************ END OF PROGRAM *************************
```

Figure 8.6—Listing for the WSH PREDICT.BAS program (continued).

of customizing for your location and entering new satellite data in the sections that follow.

When you first run the program, you'll be given the date for the satellite reference data on file and a list of the operational satellites. In this case, there will be three of them: NOAA-9, NOAA-10, and NOAA-11. Press one of the number keys (1 to 3) to select one of these satellites and get things rolling. If you press Ø, the program ends. Pressing other alphanumeric keys has no effect. For the sake of comparability with all the manual calculations you have already performed, press 2 for NOAA-10.

Once you have selected a satellite, you are asked to enter a day for the readout. Press 1 for December 1, and hit the ENTER or RETURN key. Within a second or so, the information on best passes for the day appears on your screen. The data provided includes the orbit number, the time of the crossing (in an HH:MM:SS format), the crossing point (rounded to the nearest whole degree), AOS and LOS times (in an HH:MM format), and a verbal description of the track. The track description tells you whether the pass is ascending (A) or descending (D), and the path of the track. Overhead passes are labeled. Passes labeled "near overhead east," or "near overhead west" are those that fall within 5° of the nominal overhead crossing points. All other passes are simply labeled "east" or "west," depending upon the orientation of the track relative to your station.

Once these data are displayed, you can move to the next day by pressing Y, or return to the satellite-selection menu by pressing N. Use Y to select the next few days to get a feel for how much time the program saves you, and then press N for the satellite menu.

Any day of December can be selected once you have chosen a satellite. Note that there is no direct provision for going into another month. This is because a simple program like this one accumulates errors resulting from the low orbits of the TIROS/NOAA satellites, and these errors begin to get significant as we stretch out toward the end of the month. You can get data for the next month by simply entering a number beyond the end of December. December has 31 days; should you enter 32, you'll get the display information for the first day of the next month—in this case, January 1. This extension feature is provided just to tide you over if your predict data are late.

Customizing the Program

Once you have the program working as listed, save a copy for reference, then proceed to edit lines 40-80 to customize the program for your station location:

```
40   PASSWINDOW = 15
50   ASCLONG = 69
60   ASCOVERTIME = 11
70   DESLONG = 257
80   DESOVERTIME = 39
```

All of these values should be familiar to you because they are the ones we derived for our hypothetical station when looking at the subject of pass windows. PASSWINDOW is the width, in degrees, for one-half of the pass window, centered on the nominal overhead crossing point. If you narrow this window, you'll miss some of the best passes. If you widen the window to about 20, you'll include a few additional passes. Do not widen much beyond this point, or you'll be dealing with marginal passes to the east or west.

ASCLONG is the longitude for an ascending overhead pass. Substitute the value calculated for your location. ASCOVERTIME is the nominal time to overhead following the crossing. Again, substitute your values. DESLONG and DESOVERTIME are the corresponding values for descending passes and you should substitute your values.

Now you can save the program as the master copy of your working version and proceed to check out passes in December for your location. If you have previously worked out a series of passes manually, you can compare the program results against this earlier work.

Updating Satellite Data

For ease of use with all computer systems, the reference crossing data are incorporated at the end of the program in the form of data statements. Line 1830 defines the number of satellites on file, while 1850 contains a string for the date of the reference crossing. Line 1880 contains the crossing data for NOAA-9, 1890 covers NOAA-10, and 1900 covers NOAA-11. The principle advantage of this approach is that the program will run on computers not equipped with disk drives. To use this feature, you'll have to edit these lines once each month when your new predict data arrives, inserting the new dates and satellite crossing data. Save the revised working copy for use throughout the month.

If your computer system has disk drives, as most now do, it is a simple matter to rewrite the file statements to load the data from a disk file. This file would have to be updated monthly, however, so it is questionable whether the disk version would be any more convenient.

Chapter 9 includes information on how to get on the NOAA mailing list for a Predict card, or, better yet, how to obtain the data from one of the many electronic bulletin-board systems.

The *WSH Program Disk*, available from the ARRL, was discussed in Chapter 7 in terms of the WSH1700 program for control of the scan converter. This program disk also contains a more-sophisticated tracking program for IBM PCs and compatibles. When you first run the program, data on your station location is requested and stored in a disk file. Another disk file is constructed based on information from your first Predict card. Because it's a compiled program, it runs very fast. The program itself calculates best-pass windows and other data that had to be entered manually into the simpler program presented here. While still simpler than the more-elaborate tracking programs, the disk-based program is very easy to use and is sufficient for stations using an omnidirectional antenna system.

Long-term Accuracy

If the satellite were in an absolute vacuum, orbiting around a perfectly spherical earth with absolutely uniform internal mass distribution (and there was nothing else going on in the universe), this simple program would produce results that could be used for a very long time after the reference crossing. Unfortunately, none of these conditions are true. The result is that the period of the satellite is constantly changing with a *general* trend toward a shortening of the period as the satellite's orbit gradually decays. It is these changes in the period that cause errors to build up in longer-term predictions because the period errors accumulate arithmetically with every orbit. If our orbital value is off by 0.001 minutes (less than a tenth of a second) by the end of a month, our crossing time can be in error by almost half a minute! If the error is in the order of 0.01 minutes, we'll be over four minutes off target by month's end! The major causes of such errors are the natural perturbations of the satellite's orbit and the precision of the initial orbital value that we start with in our program. We can't do much about the former, but we can attack the latter.

A graphic case of the impact of such errors arose when I was testing the program we have just discussed. It is a very simple program (the one I normally use has a listing that is over 14 pages long!), but there were errors by month's end that simply seemed too great. There are two ways to benchmark a program such as the one presented here. The first is by careful analysis of the imagery to determine the *actual* time the satellite subpoint reaches a known latitude, then backtrack to calculate what the crossing time must have been. The second approach, which I was using, was to compare the results generated by this simple program with a more-complex program of known accuracy.

There are a number of superb tracking programs available and one excellent one, SAT TRAK (written by William N. Barker and David G. Cooke and modified by T.S. Kelso), is available—along with many others—on the DRIG BBS, which I'll discuss in Chapter 9. This program uses Keplerian orbital elements (that are constantly updated and distributed by NASA) in conjunction with sophisticated algorithms that even take into account the elevation of your ground station.

The information presented by such programs is very precise (particularly if you are using the most recent set of orbital elements), and SAT TRAK can be used (in conjunction with a powerful telescope), to visually track satellites when lighting conditions are suitable. If you need precision tracking data, such a program is ideal, but it is more complex to use and (given the detail provided in the printouts) has a tendency to use up computer paper. PREDICT, on the other hand, is designed to be very easy to use, with the aim of generating primarily the data needed to make optimum use of an omnidirectional antenna system. In any case, when SAT TRAK was used to cross-check the results generated by PREDICT for December 26, the following results were obtained:

	NOAA-9	
	AOS	LOS
PREDICT (NOAA data)	10:46	11:00
SAT TRAK (NASA data)	10:44	10:58

	NOAA-10	
	AOS	LOS
PREDICT (NOAA data)	00:23	00:42
SAT TRAK (NASA data)	00:25	00:39

	NOAA-11	
	AOS	LOS
PREDICT (NOAA data)	18:37	18:51
SAT TRAK (NASA data)	18:38	18:52

Just for the record, the various satellites arrived right on schedule in terms of the SAT TRAK data. My problem was how to deal with the errors in PREDICT's listing. Results varied with individual satellites (to be expected), but the bottom line was that PREDICT was two minutes late with NOAA-9, one minute early with NOAA-10, and one minute early with NOAA-11. These errors weren't monumental, but they are great enough to mess up a tracking exercise. The crossing-point calculations were quite accurate. The NOAA-10 pass, for example, was printed out as an ascending overhead pass. In the SAT TRAK printout, maximum satellite elevation was 86.6°. Considering that PREDICT rounds crossing data to the nearest degree when evaluating the track, the results were quite acceptable!

Once math errors had been ruled out, the only possible variable was the value for the period. The following period values were contained in the NOAA Predict Bulletin for December 1, 1989:

NOAA-9	102.0213 min
NOAA-10	101.2340 min
NOAA-11	102.0851 min

I was immediately suspicious of the value for NOAA-10 because a check of past prediction bulletins showed that this value had not changed in several months. Several phone calls to the satellite service verified that

they had been using a new procedure for transferring period data from the voluminous NASA printouts they receive; the result was small, but significant, period errors. They have now adopted a different procedure that should minimize the problem in the future. Our problem, if we use the monthly bulletins, is how to determine if the period supplied is actually the best one for our use. It turns out that there are two relatively easy ways to derive a high-precision value for the period that can improve our calculations immensely.

The first scheme makes use of the two-line NASA Keplerian element sets that are available (and updated several times each month) on bulletin boards such as the DRIG system. The format for these element sets looks cryptic simply because an incredible amount of information is packed into a limited printout. An example of one of these element sets for NOAA-10 is shown below:

NOAA-10
1 16969U 86 73 A 89353.40645465 .00000892 00000-0
40960-3 0 3183
2 16969 98.6208 20.2379 0014656 72.1749 288.1026
14.23364909169038

I have underlined an item in line 1, and another in line 2. These are not underlined in the actual data set, but I did this to make it easier to point out the two items of interest when discussing the PREDICT program. There are excellent tutorials available on the DRIG BBS for decoding all of the information in the element set but, for our purposes, we'll concentrate on the two underlined items.

The item on line 1 represents the Julian day for this particular data—the December 19. The preceding digits (89) represent the year. The underlined item on the second line tells us that NOAA-10 makes 14.23364909 orbits on that day. If we divide the number of minutes in a day (1440) by this value, we should end up with a period:

$$1440/14.23364909 = 101.16871583$$

While this *is* a period value, it represents what is known as the *anomalistic period*. To convert the anomalistic period to our more familiar nodal period, we multiply the anomalistic value by 1.0056:

$$101.16871583 \times 1.00056 = 101.22537031$$

Note that this value for the nodal period is different than the 101.2340 minutes in the original NOAA predict listing. If this "refined" value, as well as comparable values for the other two satellites, is plugged into the program in place of the NOAA values, we get the following:

	NOAA-9	
	AOS	LOS
PREDICT (refined data)	10:44	20:58
SAT TRAK (NASA data)	10:44	10:58

	NOAA-10	
	AOS	LOS
PREDICT (refined data)	00:24	00:38
SAT TRAK (NASA data)	00:25	00:39

	NOAA-11	
	AOS	LOS
PREDICT (refined data)	18:38	18:52
SAT TRAK (NASA data)	18:38	18:52

Note that most of the errors between the two sets of results have disappeared except for a one-minute offset on the NOAA-10 data. Given the fact that PREDICT rounds AOS time calculations to the nearest minute, a one-minute error is about the maximum that will be encountered. If you set your timer for one minute earlier and later than the calculated AOS and LOS times, you'll probably never miss anything of consequence.

There is a second source of more refined period data: That is the periodic NASA TBUS bulletins that are transmitted daily on world meteorological HF frequencies, on WEFAX, and which are available on some bulletin boards. TBUS messages are transmitted in four parts for each satellite. The one that concerns us is Part IV, which has the following format:

PART IV
1979 057A 09345 105066560150 810414203210007 1509616
01011681 01012254 00124732 17142454 13773458 09867899
18869440 07189253 M053313427 P048448725 M000019396
P00759127 P00825300 P07350534 003263350 245206018 9449
0000499998 M00290091 P00098722 P00512415 SPARESPARE

The nodal period of the satellite is always the second data group on line two, which I have underlined for clarity. This particular TBUS abstract is for NOAA-10. Note the nodal value of 01012254, which, not surprisingly, translates to 101.2254 minutes! This is precisely the same value we would have obtained in our earlier calculations had we rounded the value to four decimal places, as is the case with TBUS nodal period values.

Other Tracking Programs

The PREDICT program listed earlier is intended to be quick and convenient and is optimized for operation with the omnidirectional antenna system. If you need precision tracking data, there are a number of useful programs that work directly from the NASA Keplerian element sets. SAT TRAK, which has already been noted, is one example. A number of public-domain and shareware programs are available on bulletin boards. AMSAT (The Radio Amateur Satellite Corporation, an organization devoted to amateur satellite communications) has a number of fine programs. The more sophisticated programs provide real-time readout for antenna bearings on your display, while plotting the satellite position on an on-screen map. Commercial programs for many different computers are also advertised in major Amateur Radio journals such as *QST, CQ,* and *73 Magazine.* Whatever your tracking needs, you should be able to find a program for your computer system.

GEOSTATIONARY-ANTENNA BEARINGS

One of the major advantages of the geostationary satellites is that, barring the movements of any satellite to compensate for the loss of another operational satellite, once your antenna azimuth and elevation have been set, they need not be changed. In effect, its like having your TV antenna permanently aligned on your local TV station.

There are various ways to determine the proper alignment, but there is no need to be too elaborate because we won't be doing the job often, and we only need to get close enough to simply hear the satellite on the first try. As long as we can hear the satellite signal after making preliminary antenna adjustments, final fine-tuning of elevation and azimuth can be done by listening to the signal on the station receiver.

We can accomplish the job rather quickly by using our orbital-plotting board. Locate the position on the equator of the satellite you want to hear. Using the edge of a piece of paper as a straight edge, locate the straight line that intersects both your station location and the satellite subpoint. Your antenna azimuth can now be read off the bearing circle normally used for tracking.

With the paper still in place, use a pencil to mark the paper at the satellite subpoint and at your location. Now shift the paper so the subpoint is still on the equator, but the paper edge passes through the pole on the diagram. Now, read the latitude of the point on the paper edge that marked your station location. This value, in degrees (N or S is irrelevant) represents the *Great Circle arc length* (in degrees) between your station and the satellite.

With this value in hand, refer to Figure 8.7. This is a plot of the relationship between the arc distance (in degrees) and the required antenna elevation relative to the horizon. Locate your arc distance on the vertical scale, then move horizontally to the right until you intersect the graph line. If you drop straight down to the horizontal scale from this point, you can read off the required antenna elevation angle.

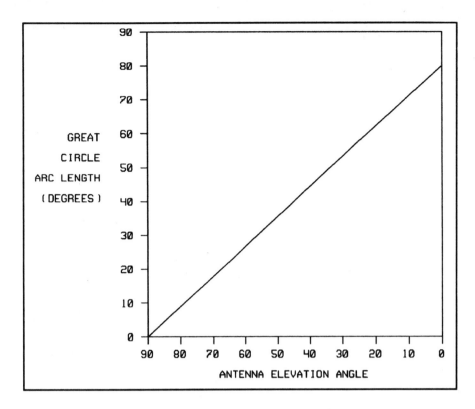

Figure 8.7—Relationship of Great Circle arc distance and geostationary antenna elevation.

Note that if the arc length exceeds 80°, the satellite will be below your RF horizon and thus out of range. Actually, any value greater than 70° arc length will put the antenna within 10° of the horizon with weaker signal levels, and raises the potential for full or partial masking of the signal by trees, buildings, hills, etc. It's a good idea to check out the bearings for satellites within range of your station before making a decision on the permanent location for the antenna. If in doubt, smaller antennas are easily mounted on a small A frame of lumber, so you can run actual reception checks from different locations.

Assuming the size of your antenna is in the 4- to 6-foot (1.2- to 1.8-m) range, you should be able to hear the satellite when the antenna is set to the indicated azimuth and elevation values. Remember that azimuth is given in degrees true bearing so if you use a compass, be sure to correct for your local magnetic declination. Many TVRO satellite dealers have bubble protractors to lend or rent that will help you set the antenna's elevation. A simple protractor and plumb bob, referenced to the feed-horn support mast, can also be used. If nothing is heard, double-check the satellite transmission schedule to make sure that the satellite is actually on the air. For larger antennas, you may have to tweak azimuth and elevation to get the first signal indication.

Once you have any kind of signal from the satellite, optimize the azimuth and elevation positions, and antenna polarization for optimum signal. The foregoing discussion assumes the use of the common az/el mounting system. If you are using a TVRO dish, you may have a polar mount. If this is aligned following the manufacturers instructions, you can use the position drive system to swing across the entire geostationary arc, zeroing in on each satellite by adjusting for maximum signal levels. If you take this approach, set your feed horn for vertical polarization, with the dish facing south, and no further polarization adjustments should be required. Once the satellites within range have been located, you can program their positions into the TVRO dish-position controller.

If you have a TVRO installation, you can often mount the S-band feed horn (facing the dish center) off to one side or the other of the TVRO feed. Do not mount the horn above or below the feed as this would require that the dish elevation be changed—a most undesirable option once the polar mount has been set up!

Chapter 9

Station Operations

SETTING UP YOUR STATION

In the "good old days," a weather-satellite station might take up a fair amount of space, what with converted surplus receivers, impressive surplus fax recorders and bulky power supplies. Today's equipment is far more compact and finding the room necessary for your equipment is less of a problem.

Certainly, your installation will be highly individualistic, but the key to planning how it should be set up is *functionality*. Ideally, every piece of equipment should be within easy reach. Obviously, while your station is still in the development stage, the gear will be spread out a bit more, if only to make it easier to work on individual pieces of equipment. Once everything is up and running, you can optimize the arrangement for your day-to-day operations. Don't shrink things too far, however, or you may find that you cannot accommodate new equipment or experimental setups.

Audio-switching units are useful for interconnecting various pieces of equipment, and offer flexibility when performing temporary tests or adding new equipment without having to pull out every piece of gear to get at the cables. You may want to install provisions for switching 12 V dc from a master power supply to the various pieces of equipment.

Safety

One aspect that cannot be overemphasized is safety. Given the modest power drain of modern equipment, the entire station can be powered from a single power strip that is protected by a circuit breaker. Use power strips with built-in surge protection, or add a surge protector at the wall outlet into which the cord from the power strip in inserted. The use of three-wire or polarized mains plugs is mandatory, and your mains outlets should be three-wire grounded receptacles.

Your station should also have a good ground connection, with heavy-gauge wire run to an outside ground rod, or to a cold-water pipe. If your home uses any PVC pipe, be sure that your cold-water-pipe ground is actually grounded. All equipment grounds should also connect to the master station ground bus. This is a major safety plus and is important to minimize RF noise from computers and other possible noise sources.

Tape Recording

An advantage of the various satellite-signal formats is that we can use standard audio-tape equipment to record the signal information for later playback. The results obtained, however, can vary from the horrible to the sublime. In years past, it was customary to blame problems associated with taped signals on the quality of the audio-tape deck. I am now forced to the conclusion that problems with taped pictures are more likely to be caused by poor *tape quality* and not the *tape deck*. Most of the images that illustrate this book were displayed from signals taped using an $80 tape deck from Radio Shack. With modern stereo cassette recorders, you can usually be assured that almost any deck can perform well in satellite service when new. A few buying guidelines are in order, however.

First, you should start with a new or almost new tape deck. Older hardware doesn't have modern enhancements, and will almost certainly be dirty and out of alignment. When shopping, look for tape decks that are intended for use in component audio systems. At any good discount electronics outlet, excellent units can be obtained in the $100 to $250 price range. Avoid dual-cassette units—for the same price, you can assume a single-cassette model has better quality. Also, watch sale prices. A more-expensive unit can often be obtained at a bargain price, especially when newer models are being introduced.

Regardless of the deck you use, keep it clean, and it will yield consistent results. The really important factor is the tape you choose and how you use it. Never use bargain tape! Spend your money on the best tape you can afford. What you are buying is a better-quality oxide, greater uniformity in its application, and higher-quality cassette hardware, all of which relate directly to image quality.

Everything else being equal, you should use 60-minute cassettes. This allows you to record two polar-

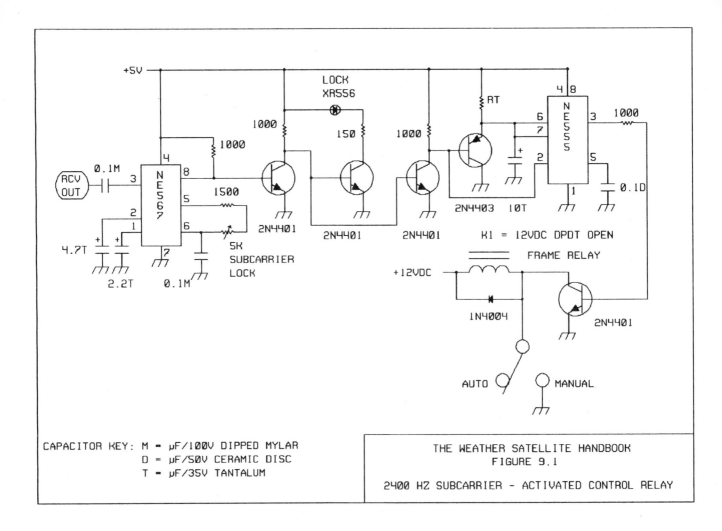

Figure 9.1—This circuit is activated by the 2400-Hz satellite subcarrier output of the receiver and features a time-delay function in the relay drop-out to minimize cycling at the beginning and end of a pass. Adjust the SUBCARRIER LOCK control to light the LOCK LED when the subcarrier signal is present at the input. The value of RT controls the time required for the relay to drop out (relay open = tape deck off) with the loss of the subcarrier signal. A value of 3.3 megohms provides a 30 to 40-second delay suitable for recording polar-orbiter passes, while a value of 470 kilohms produces a 5-second delay suitable for recording WEFAX signals. The DPDT contacts of K1 can be used to switch both sides of the ac line. Use a three-conductor (safety-ground) extension to provide power to the tape deck. All exposed ac-voltage points should be covered to prevent any possibility of operator contact during adjustment or operation. An SPST toggle switch can be included to ground the collector side of K1 to permit manual control of power to the tape deck when the switch is closed. When this circuit is used in conjunction with a timer (as described in the text), unattended recording of best passes is possible without detailed time settings.

orbiter passes per side. The 90- and 120-minute cassettes typically use thinner tape, which creates problems for the pinch rollers and capstans, and degrades dimensional stability. I have found Sony UX-S60 cassettes yield excellent and consistent performance, but other brands of comparable quality are worth investigating.

If you want to re-use a tape, don't simply record over the old material. Buy a bulk tape eraser ($20-$30) and use it to erase the old material prior to re-recording.

No single factor is as important in later recordings. Also, do not expect unlimited service life from the tapes. Even the best tape shows increasing imperfections after half a dozen recordings.

One very critical recording factor is the relative level of the video signal on the right channel to that of the clock signal on the left channel. Cassette stereo recorders have left and right tracks that are adjacent to one another; cross talk between the two channels can result if both signals are recorded at a high level.

A little video riding on the clock channel has no real effect, but the 2048-Hz clock signal bleeding onto the video track can have unfortunate consequences. The 2048-Hz clock signal can beat with the 2400-Hz video signal, modulating the video signal at a 352-Hz rate (2400 – 2048). The result is a vertical pattern of 88 light and dark lines across the WEFAX or APT frame (166 in the case of 120-LPM Meteor images). The key is to record the video signal at the highest practical level (signal peaks at 0 dB on the recording level meter). In contrast, the clock signal should be recorded at the lowest level that yields a consistent lock. I find that if I record the clock signal at a peak level of –12 to –15 dB, no visible cross talk is seen. Higher clock-signal levels result in increasingly severe beat effects.

When digital analog taping systems (DAT) become more widely available (and drop in price), our recording problems will disappear. In the interim, we must work in an analog world. Survival, in terms of good recorded image quality, equates to using a clean tape deck and the best tape you can afford.

Automatic Recording

Digital appliance timers are available that permit the tape deck to be turned on and off at selected times with extreme accuracy. Some models even let you do this several times a day with a unique schedule for each day of the week! I have been using a Radio Shack Micronta seven-day programmable timer (RS 63-889) for several months now, and find it to be quite satisfactory. Using these timers is quite simple. Program the timer to go on at AOS and off at LOS, set the deck to record, make sure the receiver is on the proper frequency, and you are set. A second timer can be set up with an external relay to switch the receiver frequency at specific times, expanding the scope of the satellites you can receive in this fashion.

You can even set up your system to receive certain satellites without bothering to program specifically when the satellite will come into range. For instance, let's say that the best-pass window (see Chapter 8) for a specific satellite falls between 0900 and 1100 hours at your location. Using the tone-operated relay circuit shown in Figure 9.1, a simple timer turns the relay circuit on at 0900 and off at 1100. The relay circuit samples the output of the receiver and turns on the recorder (the tape deck must be preset to record) whenever a 2400-Hz tone is present. For polar-orbiter service, a hold-in delay of 30 seconds or so ensures that the recorder won't cycle excessively early and late in a pass. The same timer, with a five-second delay value, can be used to record GOES transmissions. All transmissions within a window defined by the external timer will be taped with a uniform five-second interval between images. This is the ideal way to get the prime

visible-light quadrants during the day, or any other specific products you are interested in.

By using circuits such as the A-Bus system (see Chapter 7) in conjunction with a computer, you can elaborate your system as much as you want, switching frequencies and turning the recorder on and off at any time desired using the real-time clock features of your computer. You can even integrate tracking and control functions so that the computer calculates the proper times and selects the right frequency without your intervention. Add some coded tones on one channel or the other, or add a voice synthesizer, and it becomes possible to label each recording as to time, frequency, orbit number, satellite, etc. Here is an area where you can experiment using a single hardware investment and simple BASIC programming to achieve very great flexibility to capture passes when you aren't around to operate the station. Add a programmable az-el antenna mount (see Chapter 2), and the system can even track under the control of a master program.

OPERATING THE SCAN CONVERTER

When power is applied to the scan converter, the TV monitor shows a pseudorandom pattern, representing the contents of the display RAM at power-up, for about one second. The system then displays two cycles of an eight-step gray scale, interrupted across the center of the screen by a white banner strip with **METSAT 1700** in black letters. The gray-scale display can be used to fine-tune the monitor brightness and contrast controls, if required. The blackest gray-scale steps should be black, while the lightest steps should be a clean white, but without blooming or trace distortion.

WEFAX Display

WEFAX display on the scan converter is extremely simple because all functions are completely automatic. Simply cycle the **MODE** switch from **HOLD** to **WEFAX**, and display will begin with the start of the next WEFAX transmission. The display of an incoming WEFAX image involves several operations performed by the microcontroller:

- Start-tone detection
- Start-tone delay
- Phasing
- Phase-interval delay
- Image display
- Image stop

Start-Tone Detection

The scan converter software is constantly evaluating incoming video levels, looking for a pattern that is consistent with a 300-Hz start tone. This requires that the **CONTRAST** control be properly adjusted within rather wide limits. If setting the system up for the first time, it is best if you begin with manual display using the manual 240-LPM mode. Once the **CONTRAST** control has been set for good image contrast, you can use the WEFAX mode for automatic operation. The start-tone-recognition criteria are quite stringent, so the system will rarely—if ever—generate a false start, although it does recognize a start tone even with noise levels that severely degrade the quality of the final image.

Start-Tone Delay

Automatic phasing relies on looking for the black phasing pulse against an otherwise white background. Because the remainder of the start-tone waveform involves white-to-black transitions, it is necessary to bypass the start tone to avoid false phasing.

Phasing

The start-tone delay ends about one second into the phasing interval, at which point the system begins to evaluate video levels, looking for a black-level phasing pulse. To avoid false-phasing on noise pulses, the software is set up to require that a valid pulse be at least 5 ms long.

Phase-Interval Delay

Once a phasing pulse has been recognized, the scan converter's precise timekeeping ensures that later image display remains in phase. Because phasing rarely requires more than one line of the phase interval, displaying the rest of the phase interval would simply waste memory. Once proper phasing is achieved, the system waits for the image-display header.

Image Display

All of this foregoing activity is transparent to the viewer because nothing appears to be happening. The delay between the start tone and the beginning of image display depends on which GOES signal is being received. Western-hemisphere customers are sent a ROM with delay constants appropriate for the US GOES WEFAX format. In this case, display begins approximately 25 seconds from the beginning of the start tone. Customers in Europe, Africa, Asia, and the western Pacific get ROMs with METEOSAT/GMS delay values. These individuals will see a display beginning approximately seven seconds after the beginning of the start tone.

Read out of the image begins from the top of the screen, scrolling downward. You can make adjustments with the **CONTRAST** control if desired, but the uniformity of WEFAX broadcasts means that this is rarely necessary once you have obtained an optimum setting.

Image Stop

The scan converter displays every third image line until 768 image lines have been received, a process that requires 192 seconds. Then, the display stops and the system begins looking for the next start tone. Because the standard WEFAX frame is composed of 800 image lines, display normally stops about eight seconds *before* you hear the image stop tone. Some WEFAX charts are a bit longer, while tropical quadrants are shorter than normal. For such quads, you'll see the start tone at the bottom of the display, followed by a black interval at the bottom of the screen.

Once the end of an image has been reached, if you take no action, the system waits for the start of the next image and repeats the sequence described earlier. Thus, the new image is painted over the old image. To freeze an image on the display, simply cycle the **MODE** switch to **HOLD**. The image will remain on the screen as long as power is applied, or until you select a mode with the **MODE** switch.

Polar-Orbiter Display

The remaining scan-converter modes (**APT** for TIROS/NOAA visible or IR display, **240 LPM** for manual WEFAX or 240-LPM COSMOS display, or **120 LPM** for Meteor or side-by-side visible/IR NOAA display) all operate similarly. Basic operating procedures are the same for live and recorded images. Working from tapes is more convenient, however, because you can display multiple segments from the same pass, or repeat the display if something has gone amiss. Because the scan converter allows you to record the signal while you are also displaying the real-time receiver output, you can enjoy the best of both worlds.

One decision that must be made is which portion of a pass to display. For example, in the APT or 120-LPM modes, the scan converter displays a little over six minutes of image transmission, yet a typical pass yields data for anywhere from 12 to 14 minutes. If you want to center the display as much as possible on your location (assuming you're dealing with your best-pass window), you want to initiate display approximately three minutes prior to the nominal overhead time. Similarly, you can center the image display on any portion of the total pass by simply deciding when to initiate the display. Obviously, if the pass is on tape, you can display and photograph the entire pass in overlapping segments, if desired. Irrespective of such timing considerations, display can be broken down into three functional segments:

- Pre-phase display
- Phasing
- Image display

Pre-phase Display

Once you set the **MODE** switch to one of the manual display modes, display begins immediately, and the incoming data scrolls down the screen at a fairly rapid rate. This is because the scan converter is displaying every incoming image line to permit faster phasing. The scan converter is *not* passing this information to its own memory at this time and, because every line is being displayed, all features will be stretched vertically by a factor of three. If you take no action, the display simply begins again at the top of the screen once the screen is filled. (You may choose to wait a while, either to allow signal quality to improve, or simply to await the appearance of a landmark or a specific time interval.)

This continuous scrolling of the screen proceeds indefinitely until you use the **PHASE** switch or return the **MODE** switch to the **HOLD** position. In the latter case, you can then re-display the last image if desired, because no new data will have been added to the microcontroller RAM.

Phasing

Phasing is used to properly align the image and to select either the visible-light or IR displays of the TIROS/NOAA APT mode. When you press and hold the **PHASE** switch, the image begins to shift to the left as the image reads out. When the left edge of the image (see Chapter 4) reaches the left edge of the display screen, release the **PHASE** switch. At this point, the read out will seem to stop. What is really happening is that the system has reset the write-address counters, and the actual image display begins—properly phased—from the top of the screen. From this point on, the image data are also being passed to the microcontroller RAM as well as to the display screen.

Image Display

The **CONTRAST** control can be used to optimize the image as display proceeds, although you may prefer to do such touching up (if needed) during the pre-phase display. When the image display reaches the bottom of the screen, the system freezes the image and nothing more happens until you return the **MODE** switch to **HOLD**.

Post-Display Options

Once you have a WEFAX or polar-orbiter image on display and the **MODE** switch is in the **HOLD** position, several options are available. If you want to make sure that your monitor is optimally adjusted, the rear apron **RESET** switch can be used to post the gray scale that was displayed at system start up. This won't alter the image stored in the scan-converter memory and (once you have made any monitor adjustments) the image can

be redisplayed by pressing the **DISPLAY** switch. The process of redisplaying the image in RAM requires less than one second.

Ascending polar-orbiter passes (originating in the south and ending to the north) appear inverted on the monitor. WEFAX weather charts are also transmitted inverted. To correct either situation, simply press the **INVERT** switch, and the image is redisplayed as if you had rotated it 180 degrees. This process requires about three seconds. If you want to recover the original image, simply press the **DISPLAY** switch.

As this book is being written, the Soviets have been experimenting with a new 120-LPM IR format with their Meteor satellites (MET-3/3). The images are excellent, but the video format is inverted: There is an inversion of the black and white ends of the gray scale. Current versions of the 1700 ROM have been modified to correct this situation by the use of an image-complement routine invoked by pressing the **PHASE** switch. Normally, **PHASE** is ignored when the **MODE** switch is in the **HOLD** position, but with the new ROM, pressing the **PHASE** switch while in **HOLD** displays a complemented image. For standard imagery, the result looks like a photographic negative. With the new 120-LPM IR format, the result is a display of a more comprehensible image. The original version can be redisplayed at any time by simply using the **DISPLAY** switch.

HARD COPY

One of the advantages of scan conversion is that, once you have made your basic hardware investment, you can watch any number of pictures without additional expense. A CRT display requires that each image be photographed, a negative processed, and a print made to see the final results. The costs are not insignificant:

Film cost per frame:	$0.15
5 × 7 paper:	0.24
Chemicals:	0.10
	————
Total cost per print	$0.49

This does not, of course, include the value of your time (or someone else's) in doing the processing. You can, of course, use Polaroid instant film, but the price you pay will be about $0.75 per image, with a maximum print size of 4 × 5 inches. Time and economics will definitely limit the number of pictures you look at.

Fax recorders print the image directly, either on photographic paper, or on specialized electrostatic or electrolytic papers. Photographic-systems costs average about the same per image as those of the CRT system because (even though you don't have the film-processing step) you'll be using larger paper sizes.

Electrostatic or electrolytic printouts may cost as little as $0.10 per image, but the image quality is not high, and some processes are not completely permanent.

Although the use of a scan converter can eliminate the cost of viewing pictures, saving pictures does cost *something* in terms of the cost of the tape or disks used to save the image. For example, I can store 10 of the 512×480 images you see throughout this book on a single 1.44-Mbyte, 3½-inch floppy disk. Because the cost per disk is about $4, I have spent about $0.40 per image for the many hundreds of images I have stored on disk files. Lower-capacity media cost less, but the per-image storage cost won't vary greatly because the number of images you can store on a disk is also reduced.

Even with the advantages of disk storage and recall, there is always the need to be able to produce some sort of hard copy. (The production of this book is one such case.) Our two viable paths to get from a digitally stored picture to a permanent print involve either photography or various approaches to the use of a computer printer.

Photography

Getting good photographs from your TV monitor or high-resolution computer display involves a few tricks, but the process isn't difficult. First, you'll need a good camera. A basic 35-mm, single-lens reflex (SLR) camera is the best overall choice. Most standard 50-mm lenses allow the camera to focus on a 12-inch screen, and many macro lenses allow you to fill the frame with a smaller monitor screen. You want to fill as much of the image area as possible so you can produce crisp enlargements without worrying about the grain of the film.

One key to taking good photographs is to keep in mind that it takes a finite length of time for the monitor to paint the image on the screen—basically, about ¹⁄₆₀th of a second. Any shutter speed faster than this will show a partial scan! The situation is further complicated because of the nature of the focal plane shutter system used in typical SLR cameras. The shutter acts like a window shade, snapping out of the way to expose the film, then snapping back into place to interrupt the light from the lens.

As the electron beam moves during the course of an image scan of the monitor screen, the interaction of the raster scanning and shutter position will produce a diagonal pattern on the screen. To avoid this, you must make the total exposure long compared to a single scan of the screen, so that the overall effect becomes unnoticeable. I suggest exposure times of ½ to 1 second. An exposure time of this length mandates the use of a tripod, or other camera mount, to hold the camera steady during the exposure period. Also, room lighting should be very subdued so ambi-ent light does not wash out the image and reduce picture contrast. The room doesn't have to be completely dark, but it should be dark enough that the display screen looks quite bright to the eye.

Because the exposure time is set by the problems inherent in focal-plane shutter systems, the only other variable is the lens opening, or f-stop. Kodak Plus-X panchromatic film is an excellent film that offers good speed (ASA 125/DIN 22) and fine grain. Using this film, with an exposure of ½ second, the required f-stop is usually in the range of f16 to f8. With both my TV monitor and the VGA display of my computer, I've achieved good image contrast using f11. You can get a close approximation of the required f-stop by taking an exposure reading from a full-screen gray-scale display or, alternatively, you can take a series of test shots of several pictures using a range of f-stops, recording the lens opening for each shot. The negative or prints from this test roll can then be used to pick the optimum f-stop for your display. The monitor brightness and contrast settings should also be recorded so that you achieve reproducible results with each photographic session.

All of the photographs that illustrate this book were taken with a vintage Pentax camera using Plus-X film. Films were processed in Kodak Microdol X developer (7 minutes in a standard roll-film developing tank) and fixed with Kodak Rapid Fixer. Prints were made on Kodak Kodabrome RC (resin-coated) paper (normal contrast) processed in Kodak Dektol developer and Kodak Rapid Fixer. All of my darkroom facilities are very basic (a total investment of about $200), but I prefer to process my own photographs because it gives me total control over the final product. You can use commercial photo processing, but expect the process to take about a week because most photo-finishers do comparatively little black-and-white work these days and they'll wait to run films through until they have several orders in hand.

Computer Printouts

It seems there are innumerable interfaces and software systems out there that advertise "fax" capabilities. Most of these are designed to provide hard copy using a dot-matrix printer. Such printers can do an excellent job reproducing weather charts transmitted on HF, or printing the charts transmitted on WEFAX. The reason they can function well is that a chart, or other printed material (TBUS messages, schedules, and the like), represent binary data. A given pixel is either black or white. In looking at the problems of printing an image, the printer-control program can either print a dot on the paper (for black), or not print (resulting in white). Virtually any dot-matrix printer with high-resolution graphics capability can print such material with good fidelity.

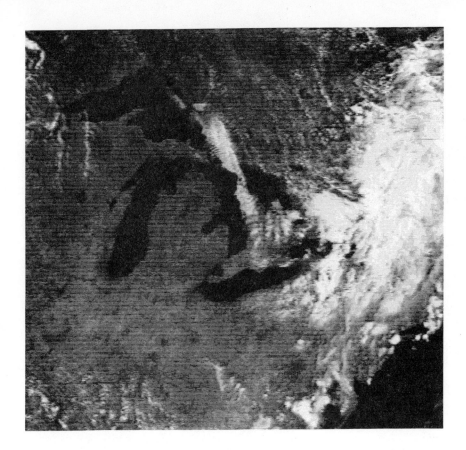

Figure 9.2—An example of a high-resolution image segment from a NOAA-11 visible-light pass. The image was captured from the scan converter streaming format using the VGA1700 program. The screen image was captured using the VGADMP screen-capture utility and printed with the PRN1700 program (see text for a discussion of both pieces of software) using a Hewlett-Packard DeskJet printer.

In order to get reasonable fidelity in a typical satellite image, we need to do far more than print simple black and white. We must be able to print varying gray-scale shades (at least 16), and do so without degrading the spatial resolution of the image. For virtually all dot-matrix printers, creation of the necessary dither patterns to reproduce gray scales takes up space on the paper. The result is, that if we have reasonable gray-scale fidelity, we end up with a picture that is lacking in spatial resolution. The result is very much like looking at a newspaper photograph using a magnifying glass. Viewed closely, such an image is textured to the extreme, and it looks good only when viewed from a distance. The images will be interesting—and even fascinating—only if you have never seen the pictures properly displayed. The typical gray-scale dot-matrix printouts are a tribute to the programmer's skill, but they are nothing more than a crude representation of the actual image.

Modern laser and ink-jet printers are another story. At relatively low resolutions of 75 or 150 dots per inch (dpi), they are no better than the typical dot-matrix printer, but at 300 dpi, using a decent driver program, they can produce results (Figure 9.2) that rival the output of many fax machines.

Making printouts with a laser printer can be expensive. First, there's the cost of the printer, the expense of toner cartridges and other supplies. Worst of all, there's the cost of anywhere from 1 to 3 Mbytes of RAM necessary to do a full page of 300-dpi graphics!

A far less-costly alternative is the Hewlett-Packard DeskJet printer. This printer does not require large amounts of RAM, and uses simple liquid-ink cartridges. With careful shopping, you can usually obtain one for something in the neighborhood of $600-$750. Although this is expensive compared to the $200 or so one might expect to pay for a 9-pin dot-matrix printer, it is not out of line with the price of some of the better 24-pin printers—and these don't even come close in terms of quality of output. Two other factors also require consideration. First, the DeskJet is almost silent. If you compare that with the incredibly annoying racket of a dot-matrix printer working in a graphics mode, there is no contest! The second additional factor is reliability. The better dot-matrix printers have a mean-time-before-failure (MTBF) rating of 4000 to 5000 hours, while the DeskJet's MTBF rating is 20,000 hours! The DeskJet is completely LaserJet compatible, so any LaserJet graphics program will drive the DeskJet. In fact, I actually prefer the output of the DeskJet in these applications.

This kind of printing requires that you have two different pieces of software. The first is a screen-capture program compatible with your display system. The purpose of such programs is to convert your

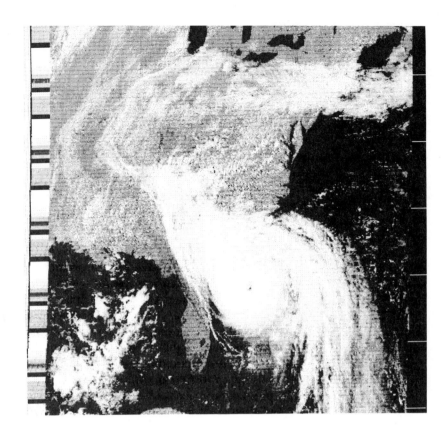

Figure 9.3—A NOAA-11 visible-light image of hurricane Hugo just hours before it pounded the city of Charleston. The image was captured using the Pizazz Plus software package and printed with the Presentation-mode display. This mode preserves the full spatial resolution of the original image, but with some gray-scale compression.

graphics screen image to a binary image file that can be used as the data source for the printing run. Such programs are typically of the terminate-and-stay-resident (TSR) type: You run the program at the beginning of an operating session and later use a set of one or more *hot keys* to initiate a screen capture. Assuming we are going to capture one of the 640 × 480 VGA screens, the resulting image file can be quite large. If a straight binary capture is used, the file size will be about 307 kbytes. If the capture program uses various data-compression routines, the file can be compressed to perhaps 150 kbytes, depending upon the image content. Once you have the image captured as a disk file, you need a printing program to provide the actual output.

There are a number of commercial pieces of software that can provide both the capture program and the printing routines. One of these is Pizazz Plus for the IBM PC, XT, AT, and PS/2. With discounting, the street price of this package is about $150. This is a very fine program that can be set up to capture virtually any type of PC-compatible screen display and print the results on almost any graphics printer in the known universe. Images can be manipulated, rotated, cropped, converted to a wide range of desktop-publishing formats, and so on.

Figure 9.3 shows an image of hurricane Hugo making landfall on the Carolina coast. This print was made using the Presentation mode of Pizazz Plus, which optimizes spatial resolution with some degradation of the gray scale. Figure 9.4 shows the same image in the Standard mode, which does a better job with the gray scale, but at some loss of spatial resolution to create the dither patterns. No attempt was made to optimize either print. The standard default settings for all functions were used. A great deal of experimenting can be done with the innumerable menu-driven features the program provides.

If there is any single theme to all of my weather-satellite work over the years, it is the idea of using your own ingenuity to get results that equal or exceed what the commercial market has to offer. Figure 9.5 should, I hope, encourage such ingenuity. This image was captured with the VGADMP shareware utility available on the DRIG bulletin board (see next section). Printing was done using the PRN1700 program that I have uploaded to this bulletin board. PRN1700 started out as a general-purpose printing program written by John Williams of the DRIG group. I modified the program to optimize it for the files produced by the VGA1700 program in conjunction with the scan converter, and further tailored the output for the DeskJet. The result faithfully reproduces the gray scale of the original and preserves the full spatial resolution of the original 512 × 480 image. Essentially the same program code has been incorporated into the VGA1700 software to per-

Figure 9.4—The same image shown in Figure 9.3, but printed with the Standard print option from the Pizazz Plus software package. There is a considerable improvement of the gray scale in this image, but at the cost of spatial resolution as the dither patterns become evident.

Figure 9.5—The same image shown in Figures 9.3 and 9.4, but captured and printed with the software tools noted in Figure 9.2 and discussed in the text. This printout optimizes both gray-scale and spatial resolution and rivals the output of many fax systems. In all three figures, the original image was derived from the scan converter using the VGA1700 program.

mit Desk/LaserJet printing directly from the program image files.

Although such programs are normally used in conjunction with high-resolution displays such as VGA, there is no reason why similar techniques couldn't be used even if you lacked a high-resolution display. It would be a fairly easy matter to capture the high-resolution data stream from the computer and use this data to create a disk file employing the same format as any screen-capture routine. The contents of this file could then be printed, and the results would be indistinguishable from an image that actually was captured from a VGA screen!

SATELLITE BULLETIN BOARDS

Once you begin to reach operational status with your station, you'll need to keep up with a veritable flood of information. Orbital elements, predict data, new satellites, the status of existing systems, hardware and software evaluations—the list seems endless. The most effective way to accomplish this is to add a modem to your computer and take advantage of the wealth of information available from BBSes.

The premier BBS service for weather-satellite enthusiasts is the one operated by the Dallas Remote Imaging Group (DRIG), spearheaded by the SYSOP, Jeff Wallach. DRIG is dedicated to providing the most up-to-date information on weather satellites, remote imaging, Amateur Radio satellites, and the space program in general. It would take a full chapter just to outline the tremendous range of resources provided by this bulletin board. The DRIG provides all the predict data you need to keep track of the US and Soviet polar orbiters (and many other satellites), and they have a FILES section that boggles the mind. Public-domain and shareware programs for satellite tracking, screen capture, high-resolution image display on all the major computer systems, and an unparalleled library of APT, HRPT, and VAS images are available. Here you'll find new products announced and reviewed, and the electronic mail section provides you the opportunity to ask questions that will be answered by some of the sharpest folks in the business. If you have gone to the trouble of putting together a satellite station and then fail to get a modem and connect to the DRIG system, you are missing out on the most current source of information available to you.

The DRIG bulletin board can be reached at 214-394-7438. Communications parameters are 8 bits, 1 stop bit, and no parity. Data rates up to 2400 bauds are supported. You can log in and download bulletins, predict data, Keplerian elements, etc, at no charge. No one will brow-beat you for a user charge, but if you do begin to use the system on a regular basis (and you will!), you should consider making the suggested donation of $24 per year that gives you Professional membership status and access to expanded privileges. It takes a tremendous amount of time and computer hardware to maintain a system like this and the donations help to ease the burden on the dedicated group of individuals who maintain the system. Donations can be mailed to:

DATALINK RBBS
PO Box 117088
Carrollton, TX 75011-7088

The DRIG system is linked to other major bulletin boards, including the NOAA EBBS system, assuring you of the most up-to-date data and providing the information that can let you branch out to other systems if your needs dictate.

NOAA operates a BBS to assist ground stations and others in the user community in keeping up with satellite news, schedules, etc. This board carries monthly predict data, Keplerian elements, WEFAX transmission schedules, and a host of technical memoranda and news notices. Much of this material is downloaded to the DRIG board every week or so. Checking into and using this system is a bit more complex than that required for the DRIG system. For this reason, I suggest that you write to NOAA at the address given in the next section and ask for information on the NOAA EEB system. They'll forward to you a pamphlet with all the particulars.

As this edition goes to press, I'm putting a 24-hour bulletin board system (the WSH BBS) on line as a service to readers of the *Weather Satellite Handbook*. (The telephone number for the BBS will be posted on the DRIG BBS.) The WSH BBS operates at 300/1200 baud, using 8 data bits, 1 stop bit, no parity. The board carries bulletins covering the scan converter, software and hardware developments, construction hints, and the current NOAA Predict data and NASA Keplerian elements for operational weather satellites. The files area contains a small number of sample images (updated regularly) and utility programs. You'll be able to leave questions or comments and receive answers to your questions. The BBS is open to all, and there's no charge for using the service.

The BBS is not a message-transfer system. You can leave questions and receive answers, but the BBS will not accept message transfers between users; this is a function best handled by larger systems. Additional features may be added at a later date, but the intention is to operate a modest system, tightly focused on the *Handbook* and related hardware and software projects. I encourage you to use the system, for replies to questions will be far more timely than using the mail. I plan to continue to maintain the system as long as there is a reasonable level of activity.

PREDICTS AND INFORMATION NOTES BY MAIL

For many years now, the mails have provided NOAA with the principal means of keeping in touch with the user community. As program complexity has increased, along with the number of user stations, the cost of preparing and mailing materials has become a serious problem. For this reason, the satellite service has opted to use the electronic bulletin board as its primary channel for disseminating data while encouraging other systems to download the data and make it available to their users. They do realize, however, that not all users have the facilities to make use of such electronic information systems. If that is your situation, you can contact the satellite service directly and request to be placed on any one of three mailing lists:

1) Monthly mailing of the APT PREDICT postcard containing the reference-crossing data for operational polar-orbiting satellites.

2) The APT Information Note list that allows you to receive mailings of information notes relevant to polar-orbiting satellites.

3) WEFAX Information Notes—current data, schedules, etc, for the operational GOES satellites.

It takes time and money to disseminate these materials by mail, so don't ask to have them provided simply out of curiosity. The material is meant to assist operational satellite stations. If you have a computer—and particularly if you are located in North America—by all means make the modest investment in a modem and tap into the electronic information systems. The information you receive will be more timely, you'll get much more information, and the cost to the already-strained direct-read-out service budget will be reduced. If mail is your only option, send requests to the following address:

Office of Direct Readout Services
NOAA/NESDIS
US Department of Commerce
Washington, DC 20233

JOURNAL OF THE ENVIRONMENTAL SATELLITE AMATEUR USERS GROUP

The *Journal of the Environmental Satellite Amateur Users Group* (JESAUG) has been published for a number of years now under several dedicated editors. The *Journal* is published quarterly and is dedicated to assisting both beginning satellite experimenters as well as documenting some of the latest advances in the amateur user community. If you are on the road to becoming a dedicated satellite enthusiast, a subscription to JESAUG is a virtual necessity. JESAUG publication has recently been assumed by the DRIG group; a subscription costs $30 per year in the US and Canada, and $40 a year elsewhere. Subscriptions to JESAUG can be obtained from:

Dallas Remote Imaging Group
JESAUG
PO Box 117088
Carrollton, TX 75011-7088
Tel 214-394-7325

IN CASE OF DIFFICULTY

Taken as a whole, the current system of weather satellites is amazingly complex: the satellites themselves, the ground-support systems, and (last, but not least) your ground station. Inevitably, there will be times when you notice problems; the question then will be where the problem originates. Your first step should be to thoroughly check the various components of your station, using several satellite signals. If you are having similar problems with a number of satellites, the likely source of the problem is in your own installation. Even if all the electronic systems are functioning, there is always the possibility of water in connectors or transmission lines, or other subtle problems that need checking.

If the problem you're experiencing seems to be with a single satellite, some other factor may be involved. Use the electronic BBSes to see if others have noted the problem or focused on solutions. Only when the problem can be definitely assigned to the satellite (or the ground-support system) should you proceed to check with the folks at NOAA. The amateur community has acquired a high degree of credibility over the years, and there is no need to compromise that by complaining about problems that cannot be verified by several sources. We provide a valuable monitoring function that can be a real service to the technical agencies that operate the satellites, but it is imperative that our reports be absolutely reliable. While you are still relatively new at this game, double-check your observations with others to avoid the embarrassment of taking up the agency's time, only to discover that you are the one with the problem. There is a huge reservoir of talent on tap on the BBSes. With all of us willing to pitch in, your problems can probably be localized and corrected in short order.

Chapter 10

Advanced Applications

INTRODUCTION

Once you have a computer interfaced to your scan converter, a whole range of new applications begins to emerge. What you can accomplish is largely a function of your imagination and programming skills. One of the fascinating aspects of upgrading your programming abilities is that it is certainly educational, it's a mix of fun and frustration, and it rarely costs a lot of money. Some of the applications I'll discuss can be accomplished in BASIC, but most require the speed of a fast, compiled language such as C or—better yet—assembler. If you begin to seriously approach new frontiers in image handling, time spent learning to program in assembler for your system is time well spent.

AUTOMATIC DIGITAL ACQUISITION AND STORAGE OF IMAGES

The conventional way to capture pictures when you are not present is to use a tape recorder (see Chapter 9). Although this works very well, as indicated by the majority of the pictures in this book, until digital audio tape (DAT) systems become widely available at a reasonable price, recorded images can never have quite the quality of "live" images.

If you have followed through to this point in the development of your system, you have all the tools you need accomplish automatic operation. Your computer can set the operating mode of the scan converter and can transfer images from the scan converter to disk. Because most computers can also tell time (using a real-time clock or the time/date set functions at log-on), it's possible to automatically capture images from satellite passes when you're not available. Starting with the WSH1700.BAS program in Appendix III, for example, you can add the following routines.

Disk-Based Pass Log

Your manual or computer-based tracking system can tell you when passes for various satellites will occur.

It's a simple matter to write a program to create a disk file containing information on the date, time, and mode (based on the satellite in question) for each of a series of passes. Because the scan converter displays six minutes of image data, the time you enter would be about three minutes *prior* to the nominal overhead time or time of closest approach.

Auto-Acquisition Module

With a little bit of additional programming, you could add a routine to the WSH1700.BAS program that would do the following:

1) Check the disk file for the next pass.
2) When the date and time in the file matches the current date and time, switch the scan converter to the appropriate automatic mode.
3) Watch the computer input for the 255 code that indicates when the image has been loaded.
4) Reset the scan converter mode
5) Initiate a transfer of the image from the 1700 to the computer memory.
6) Save the 64-kbyte image file to disk. You can derive the file name from the TIME$ is you wish, remembering to append a .WSH extension.
7) Go back to Step 1 for the next pass.

When you return, all the images logged should be on disk files and you can load them as you would any other. Because the images were captured live, they should not have any of the possible imperfections introduced by recordings. Obviously, this technique can be used to capture images from almost all of the passes of a given day if you have a scanning receiver (such as the new Vanguard unit described in Chapter 3), but you could use the unused functions of the C Port in your interface to arrange for manual switching (using reed relays) of a crystal-controlled receiver. With additional imagination, you could incorporate control routines for operation of an az-el rotator system. This would rarely be necessary because you would

A

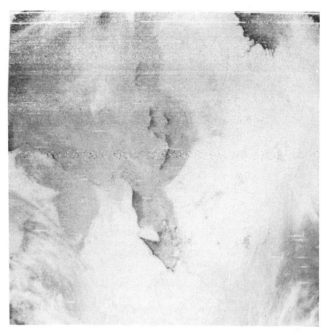

B

Figure 10.1—An example of image complementation. (A) shows a Meteor 3/3 120-LPM IR image as received. This image looks strange because the clouds are black as a result of the inverse modulation used in the IR format for this satellite. (B) shows the same image following complementation by means of reversing the order of the PALETTE statements.

normally want to capture images that cover your area, and the Zapper omnidirectional antenna (described in Chapter 2) will do a fine job.

High-Resolution Images

The preceding example will handle the images at the stand-alone resolution of the '1700, but the same basic approach can be used for the high-resolution data streamed by the scan converter during image acquisition. The VGA1700 program from Metsat Products (see the Commercial Software discussion in Chapter 7) performs just this function. The program supports the creation of a file of pass data, uses that data to initiate capture of the high-resolution image data and stores the data on disk.

Because of the high spatial resolution and the fact that 6-bit data are stored, the disk files are considerably larger than 64 kbytes. These high-resolution binary-image files can be loaded into the system and processed with the same image-enhancement features that will be described in the following section. In effect, I can capture data from any number of passes each day and have the disk files waiting for me when I get home! The end result is that I have as many options as if I had been present for each pass but, instead, I let the computer do the work while I am out trying to pay for all this innovation!

SOME IMAGE-PROCESSING BASICS

One of the marvelous aspects of digital images is that the images themselves exist entirely as a collection of numbers. Lewis Carroll, who gave us the marvels of *Alice in Wonderland*, has the caterpillar say to Alice …"words mean what I say they mean"… Our precocious insect larva might as well have been talking about digital-image processing. The numbers representing the image can be made to mean almost anything: For numbers can be changed in an amazing number of ways. When we change the numbers, we change the image. The key to digital-image processing is to change the numbers in ways that make the image more useful or interesting. There are so many possibilities that we cannot possibly cover them all. If you want a good introduction to the subject, I suggest that you get a copy of G. R. Baxes "Digital Image Processing," (Prentice-Hall, Inc, 1984, Englewood Cliffs, NJ 07632, 182 pp.). I'll touch on a few of the many possible applications, but the field is almost open-ended.

Turning Negatives into Positives

In the design of the new Soviet IR imaging system being used by Meteor 3/3, the engineers threw us a curve. For years now, weather-satellite IR data has used the white end of the dynamic range for cold objects

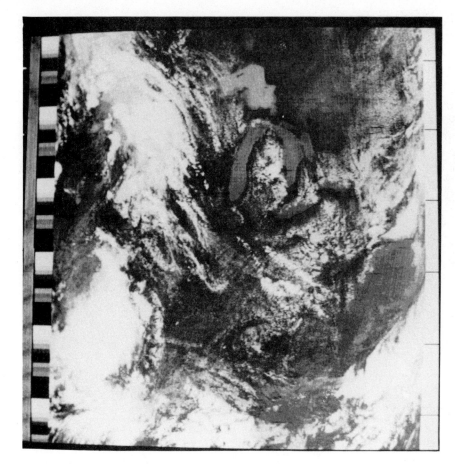

Figure 10.2—A NOAA-11 IR view of the east coast of the US in late summer. The waters of the Great Lakes are cooler (lighter) than the surrounding land surfaces. Note that southern Lake Michigan is warmer than the north. The water of Lake Superior is very cold, except for the comparatively shallow areas around the southern and eastern shores. Isle Royale, comparatively warm in mid-afternoon, shows up in strong contrast against the surrounding cold water.

and the dark end for warm. The new Meteor format reverses this relationship; the result is an image format in which the clouds appear black and the entire image has the look of a photographic negative (Figure 10.1A). Fortunately, there are several ways, all of them easy, to convert such an image to the more familiar format (Figure 10.1B).

Perhaps the fastest way is open to users of VGA displays or other high-end video systems with similar capabilities. At its highest resolution (640 × 480) the standard VGA card can display 16 color attributes, numbered Ø-15, for any pixel. Note that I used the term *attribute* rather than color or gray-scale intensity. The reason for this is that each of these 16 attributes can be assigned any color or gray-scale value over an incredibly wide range by the use of the PALETTE statement in BASIC. For normal display, PALETTE is used to assign gray-scale values in a stepwise sequence with attribute Ø assigned to black and attribute 15 set for full white. The beauty of the PALETTE statement is that a single line of BASIC code can reassign all the colors/intensities for the attributes and the change is instantly implemented throughout the image. In the case of the negative IR image, we can use PALETTE to instantly reverse the gray-scale assignments of the attributes such that attribute Ø is white and attribute 15

is black. This has the effect of instantly shifting the negative image to the more-familiar positive format. Using PALETTE to reassign the original values instantly puts us back to the "normal" display.

When I first recognized the negative nature of the new format, it took only a few minutes to add a complement feature (the mathematical equivalent of reversing negative and positive) to the source code for the VGA1700 program. This program is written in compiled BASIC (Microsoft QuickBASIC 4.5) along with an assembly language module for the operations that require great speed. A few lines of BASIC code to add the complement feature, plus another to return to normal display, recompile the code and the job was done. This is precisely the approach used to convert the image set in Figure 10.1. The conversion involves pressing a single key (<C> for complement or <N> for normal) and the change is immediate.

If the use of a PALETTE statement or its equivalent is not applicable in your case, there is still another simple technique. For example, the scan converter stores two pixels per byte in the display memory. To implement the complementation function in passing the data from the CPU RAM to the display RAM, we can use a simple mathematical approach that is universally applicable. If we subtract the actual value of a pixel from the *maximum* pixel value, the remainder represents the complemented version of the pixel. This is most easily visualized at the extremes. If we are dealing with 4-bit pixels, the maximum pixel value is 15. If we have a white pixel (value = 15) and subtract this value from 15, the remainder is $\emptyset$, or black! Similarly, if a black pixel (value = $\emptyset$) is subtracted from 15, the remainder is 15, or white! Although not as obvious, the same relationship holds for all intermediate pixel values (1-14). If we were using 8-bit pixels, we would subtract the original pixel value from 255 (the maximum value for a byte). In any system, we have the choice of complementing the memory values in RAM or doing the job when the pixel is passed to the display. Both approaches work equally well and the choice is based in your application.

False-Color Display

Our satellite images are monochromatic, but it's occasionally useful to convert the gray-scale values to specific colors. Two closely-spaced gray-scale values may be hard to differentiate, but seeing the contrast between them is easy if they are assigned to different colors. Such false-color displays are useful in emphasizing temperature differences in IR display, for example. Figure 10.2 illustrates a high-resolution NOAA IR view of the Great Lakes that shows some very interesting temperature profiles across the lake surfaces. Although these are clearly evident, they would literally jump out at you if the subtle shades of gray were converted to contrasting colors.

If you're using a color VGA display, our ever-useful **PALETTE** statement can do the job instantly. Each of the 16 display attributes can be re-assigned to specific colors and the picture will instantly shift from a black-and-white gray-scale display to living color! You can also generate some interesting color effects with NOAA visible-light imagery. For example, water usually displays as black, so we can assign attribute $\emptyset$ to a pleasing shade of blue. Darker gray scales can be assigned to a range of greens, lighter grays to browns, and the last few steps left as white-intensity values. If we implement such a palette change, the visible-light picture will show water in blue, terrain in shades of green and brown, and the clouds in white. The precise effect varies with your **CONTRAST** control setting, but you can usually create an image that looks like what you imagine a color photograph would produce.

False-color displays can look chaotic, so you won't want to use this approach for day-to-day display, but it's handy to be able to implement it for special effects.

If you like to tinker, it's a relatively simple matter to use an RGB or RGBI TTL color monitor to view false-color images directly from the scan converter. Take a look at the technical specifications for your monitor and determine what kind of interface connector is required, where the color lines and sync signals connect, and what the polarity of the sync signals might be. The color-signal lines can be taken off the output of the 74LS157 through 10-ohm resistors. Individual sync signals of high- or low-going logic are available at the 74LS221, and a composite sync signal is available at the sync mixer.

With an RGB monitor, we're dealing with binary logic levels (high or low) on three lines, each controlling whether a specific color gun is on or off. Because there are three lines, each with two possible logic states, there are eight possible color combinations (2^3), depending upon the logic patterns:

R	G	B	Display Color
L	L	L	Black (BLK)
H	L	L	Red (RED)
L	H	L	Green (GRN)
H	H	L	Yellow (YEL)
L	L	H	Blue (BLU)
H	L	H	Magenta (MAG)
L	H	H	Cyan (CYN)
H	H	H	White (WHT)

If we connect the three high-order output bits from the scan-converter 74LS157 multiplexer, we get a specific relationship between the 4-bit pixel value and the resulting screen color, depending upon which of the output bits is connected to which color line:

On an RGB monitor, there are eight colors available, and the palette will depend on which output bits are connected to which color line:

Output Bits			Video Value							
			$\emptyset$	2	4	6	8	1$\emptyset$	12	14
1	2	3	1	3	5	7	9	11	13	15
B	G	R	BLK	BLU	GRN	CYN	RED	MAG	YEL	WHT
G	B	R	BLK	GRN	BLU	CYN	RED	YEL	MAG	WHT
B	R	G	BLK	BLU	RED	MAG	GRN	CYN	YEL	WHT
G	R	B	BLK	GRN	RED	YEL	BLU	CYN	MAG	WHT
R	B	G	BLK	RED	BLU	MAG	GRN	YEL	CYN	WHT
R	G	B	BLK	RED	GRN	YEL	BLU	MAG	CYN	WHT

Note that because the lowest-order output bit ($\emptyset$) is not used, pairs of possible pixel values share the same color. If you have an RGBI monitor (where the I is an *intensity* line), we now have 2^4, or 16 possible logic combinations. With any given combination of RGB logic levels, if I is high, the colors will be brighter, or more intense, than if I is low. If we assign I to the high-order output bit (3), red to bit 2, green to bit 1, and blue to bit $\emptyset$, we would get the following color relationships:

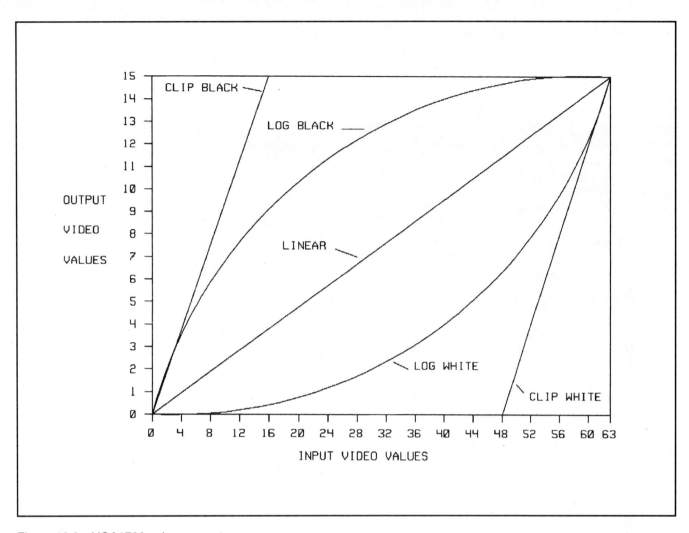

Figure 10.3—VGA1700 enhancement curves.

Video Value	Color
Ø	black
1	dark blue
2	dark green
3	dark cyan
4	dark red
5	dark magenta
6	dark yellow (orange)
7	dark gray
8	light gray
9	light blue
10	light green
11	light cyan
12	light red
13	light magenta
14	light yellow
15	white

In this case, each possible pixel value has its own unique color, but we are limited to the color sequences provided by the pattern of connection of the output bits to the color and intensity lines.

What do we do if we want to set up some other relationship between video values and display color? The colors are a given, but we can assign *any* color to *any* pixel value by means of a very useful technique known as a *look-up table*. Depending upon how our color lines are connected, each one of our 16 possible colors has a unique numerical value. We can set up a table in RAM that we use in passing video values from memory to the display.

For example, in the sequence shown earlier, a pixel value of 9 yields a light-blue display. If we wanted a video value of 2 to display as that light blue, we would check the value of the pixel in RAM and, if it were a 2, we would change it to a 9 prior to passing it to the display RAM. This can be done very easily in BASIC, but it's a slow process. In assembly language program-

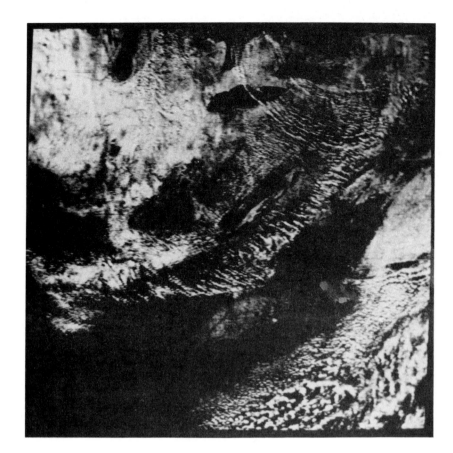

Figure 10.4—A high-resolution view of the northeastern US after a Christmas-day snow storm resulted in snow flurries as far south as Miami. The linear transform was used for this image. Although excellent cloud detail is available, land areas not covered by snow are not visible.

ming, we could arrange the desired sequence of display video values in order, starting at some base address. When we fetched a pixel value from RAM, we'd set a pointer to the first table entry, then increment the pointer by a count equal to the pixel value. The value for the display pixel could then be recovered from RAM at the new pointer location! As we shall see shortly, look-up tables are useful for other video-processing routines as well.

Video Processing

If we have more bits for each pixel in RAM than we actually need for display, we have great flexibility in what we can do with the video information. The scan converter, for example, streams 6-bit, high-resolution data which is captured by the VGA1700 software discussed earlier. This program stores 6-bit video values in RAM, yet the VGA display only accommodates 4-bit data. Each time we display an image from memory, we must transform each pixel from 6 bits to 4 bits and, in doing so, we have a very powerful tool at our disposal. Figure 10.3 shows the five default video-processing curves that are built into the VGA1700 software package. Note that each of the curves (clip black, log black, linear, log white, and clip white) defines a specific relationship between 6-bit input values (values from 0 to 63 in RAM) and the 4-bit output values (0-15) used for display.

Linear Transform

When the scan converter is loading an image, the VGA1700 software captures the pixel values streamed from the scan converter and stores then in the computer RAM in 6-bit format. At the same time, it's also converting the 6-bit values to 4-bit values to pass to the VGA display in real time. When it does so, it uses a look-up table loaded with the linear-transform values. Note that the linear curve is a straight line that maintains a linear relationship between input and output over the entire input dynamic range. The result is that the display preserves all the brightness relationships of the original 6-bit data. The linear transform is the one of choice for WEFAX data because WEFAX images tend to be optimally processed by the ground computers and there is often little to be gained by manipulating pixel values in a nonlinear fashion. The linear transform also works out well, given careful contrast adjustment, for warm-weather IR imagery from the TIROS/NOAA satellites.

Visible-light polar-orbiter imagery is another matter entirely. If we adjust the contrast so the very whitest clouds are white, we can get excellent detail in cloud structures, but the ground detail is usually minimal,

Figure 10.5—The same image data used to display Figure 10.4, but processed with the log-black transform. Had the contrast simply been increased to bring up the land areas, most cloud detail would have been lost. The log-black curve brings out the land features without losing all the structure in the cloud systems. The patch of snow cover along the Carolina coast (barely visible in Figure 10.4) is clearly evident in this view and can be related to other terrain features.

with land and water features reproducing as black or very dark grays. We can bring out these features by increasing contrast, but if we do so, we'll tend to lose all the cloud details, and all clouds will reproduce as a uniform white. To recover ground detail *and* preserve some of the detail in the clouds as well, we must use a nonlinear transform—in this case, the log-black curve.

Log-Black Transform

Note that the log-black curve is decidedly nonlinear and results in the assignment of a large number of the output pixels to the low end of the input gray scale, with comparatively fewer pixels allocated to the white end of the scale. We use this curve following the loading of a streaming image from the scan converter. The initial display will be linear and—assuming the **CONTRAST** control has been set for optimum cloud detail—ground details will be minimal.

Figure 10.4 shows a high-resolution segment covering the US east coast on the day after a Christmas freeze that saw snow flurries in Miami and snow on the ground throughout the coastal Carolinas. The clouds are well defined in this image, but ground detail is essentially lacking south of the area of heavy snow cover. When the log-black processing curve was used on this image, the results seem almost magical (Figure 10.5). We have retained some cloud structure, but

ground detail has emerged, seemingly from nowhere. Actually, the ground detail was there all along in the 6-bit data, but simply wasn't discernible when the 6-bit data were transformed in a linear fashion to 4-bit display data.

With the original data in memory, any of the available curves can be used to alter the display or the linear display can be recovered at any time. Because most of us like to see such ground detail—if only to assist in orienting the image—most of the NOAA visible-light images in this book were displayed and saved to disk using the log-black transform. The remaining default curves each have their specialized uses.

Clip-Black Transform

I have noted repeatedly in earlier chapters how difficult it is to see any ground detail in Meteor visible-light images. Even the log-black transform typically fails to show anything. In such instances, the clip-black transform can be used in conjunction with a higher **CONTRAST** control setting to recover enough detail to orient the image. Clip black assigns all the output gray-scale values to the lowest 16 input values, and any value higher than 15 is simply assigned as white. Such images are often not very appealing because of the extreme processing, but you can often see enough to orient the image.

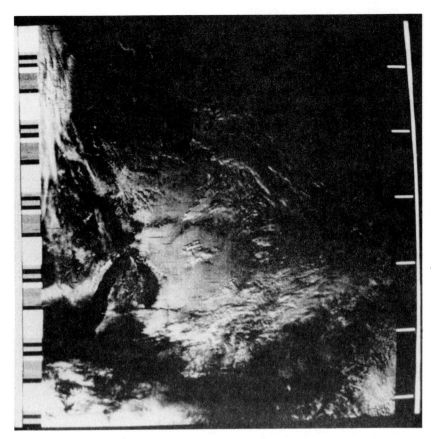

Figure 10.6—This image, a NOAA-11 afternoon pass in January, was processed with the linear transform and is very dark to the northeast; only ground features with snow cover are visible. The effect of the log-black transform, also useful for compensating for the unequal lighting characteristic of winter daylight imagery shown here, can be seen in Figure 10.7.

Figure 10.7—The log-black transform applied to the image data in Figure 10.6. Coastline and lake shores are now evident, surface and cloud detail to the north are much improved. Overall, the cloud features are displayed far better than if the CONTRAST control had simply been adjusted in an attempt to compensate for the poor lighting.

Figure 10.8—When a computer with high-resolution display capabilities is available (VGA display in this case), it's possible to sample the high-resolution data stream from the computer to produce a sub-segment of the image at essentially full resolution. This is a NOAA-11 ascending (afternoon) pass in the early fall. The picture includes an area from the eastern Great Lakes to the northeastern US coast and the Canadian Maritime provinces.

Log-White Transform

This curve does for the white end of the gray scale what the log-black transform does for the low end of the dynamic range. It provides moderate—but not excessive—contrast expansion for IR imagery.

Clip-White Transform

This is an extreme transformation routine useful with some types of winter IR imagery. The curve assigns gray-scale values to the extreme upper end of the input dynamic range, and any input value less than 48 is simply assigned to black. Like the clip-black routine, the results are rarely visually appealing, but it's often possible to recover detail that could not be observed in any other way.

In addition to the five default curves, the VGA1700 software lets you create your own custom curves and store them as disk files. Any one of these can be called up from disk and used as needed for specific situations.

Although this discussion concentrated on the use of the VGA as the display medium, any computer can capture the 6-bit streaming data to memory and use similar transform techniques in passing the data back to the scan converter for display. This feature permits

you to do video processing without a high-resolution computer display.

HIGH-RESOLUTION DISPLAY

In the streaming mode, the scan converter is passing 1024 pixels per line and 768 lines of image data—essentially the full resolution of the image format. This is roughly twice the resolution that a standard VGA display can handle, but we can take advantage of the detail inherent in the streaming data by sampling just a portion of the incoming image at full resolution and passing only that portion of the image to the display. This is the technique used for the various "high-resolution" images you'll see in this edition, and there are two previous examples in this chapter. Figure 10.8 is another example.

The VGA display may not be able to show all of the incoming image at full resolution, but it can display about one-quarter of the total image area with full detail. The VGA1700 software package supports such high-resolution sampling routines. The technique is not limited to high-resolution displays. Any computer can sub-sample the data stream and pass the results back to the scan converter for display on the TV monitor. Although not as detailed as the equivalent VGA images (see Figure 10.9), the results greatly ex-

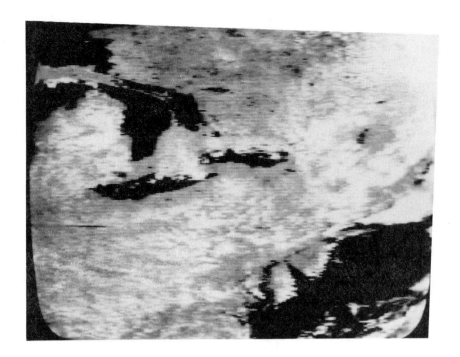

Figure 10.9—An example of the use of the scan converter to display higher-resolution data. The computer captured the high-resolution data stream from the scan converter during image loading and formatted a 32-kbyte subset of the original image that was passed to the scan converter as described earlier in this chapter. Though not as detailed as a high-resolution image displayed on the VGA system, the resulting picture has twice the resolution of the scan converter in the stand-alone mode, and does not require that the computer have a high-resolution graphics display. This image sub-segment covers the eastern Great Lakes and the US northeast coastline.

ceed what the scan converter can accomplish in the stand-alone mode. Such techniques could have been incorporated directly into the scan converter ROM software, but a much more complex front-panel switching scheme would have been required.

MOSAICS

Most of the multi-image mosaics in this book were prepared by manually joining photographs of individual display segments, but creative programming can accomplish the same job. Figure 10.10 is a sample of a computer-generated mosaic, linking the northern and southern portions of a NOAA-11 visible-light pass. Linking segments of a single polar-orbiter pass, or a set of four WEFAX quads to create an image of the entire earth disk, is a relatively simple programming task. In each of these cases, the pieces all fit at right angles, and programming is simply a matter of selecting the appropriate offsets to fit the images together. If you want a real challenge, consider the problems of merging an eastern and western polar-orbiter pass where the units must be merged at an angle (see the cover of this book). It can be done, but it's a major software challenge on a home-computer system!

SUMMARY

Although this discussion has only scratched the surface of the many possibilities inherent in teaming a computer up with the scan converter, it should be obvious how flexible your options are. Your modest investment in the scan converter continues to pay dividends as you increase the capabilities of the computer system, whatever it might be. My key consider-

ation in the design of the scan converter was to provide a piece of hardware that would meet the needs of those just getting started in this hobby, as well as more seasoned experimenters. I think it is safe to say that that objective was realized.

POSTSCRIPT

In any project, you must come to points that represent an ending of some sort. In a sense, I have to make some arbitrary decisions about where and when to bring this particular edition to a close. The decision is arbitrary in the sense that new satellites are to be launched, each day brings new and fascinating images into the station, and I cannot help starting new projects or playing with new ideas. At times like this, I hearken back to the words of my mentor in graduate school who admonished all of us that we could never write the last word: At some point, it was time to finish the work!

As I write this, a NOAA-10 pass is being painted on the TV screen. This edition of the *Handbook* is not the last word on weather satellites, either in terms of my own work or the work of thousands of other satellite experimenters who share this hobby. It will, however, get you started. Even if you go no farther, you can enjoy an activity that will provide endless fascination for years to come. It can, however, be just the beginning. For, each of the talented folks that are now at the cutting edge of new technology started with a basic satellite ground station.

When many of us began, the satellite service had a tendency to use the term "amateur" in a perjorative sense. We were basement tinkerers, and I suspect that

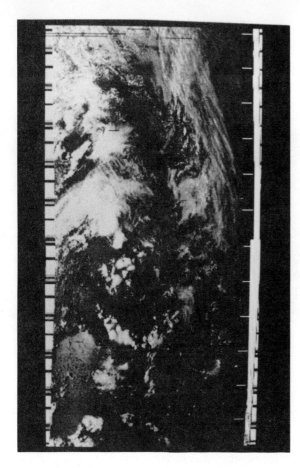

Figure 10.10—An example of a computer-generated mosaic, linking the northern and southern halves of a NOAA-11 ascending pass. The two components of the pass (each representing about six minutes of data) were displayed on the scan converter, then passed to a computer (as described in Chapter 7). The computer analyzed the last image line in the northern image, then searched the southern file for the same image. When a match was located, the computer offset by one line, then computed the byte offset required to provide the best lateral fit. The resulting composite image was then displayed on the VGA screen. The entire process took less than a minute with no operator intervention (ignoring, of course, the time invested in writing the program!).

many of the professionals felt put upon by our incessant questions and occasional criticism. Over the years, we amateurs accomplished something the professionals didn't have the time to do: We made quality weather-satellite stations affordable to the point where this immense resource created by the space program can now be a part of any classroom or home. Anyone who needs or wants to watch the earth from space can now do so. That is an achievement in which we all can take pride. If you have come this far, there is a good chance that you're ready to join our unique fraternity. Do so. You'll gain tremendous personal satisfaction and make friends for a lifetime.

Now, on this first day of January 1990, I bring this work to a close. In doing so, I trust that for both you and me, it simply marks a new beginning.

Glossary

It is impractical for me to hope to define every term with which you may not be familiar. This glossary includes those terms that are most relevant to satellite-station operations, but largely excludes terms used in general electronics and radio technology—including most antenna terms. A good general reference covering most of these areas is *The ARRL Handbook*, published annually by the American Radio Relay League.

Active filter—an amplifier (usually an op amp IC), used in conjunction with specific resistor and capacitor values to create a filter with desired characteristics (see Filter).

Address—combination of logic values on an address bus that specifies the location of a particular byte of data in a memory chip.

Address bus—the combination of individual address lines in a microprocessor system that, in conjunction with other signals, defines memory locations, I/O ports, or other elements in the memory map.

AM—*Amplitude Modulation*. A method of placing information on an RF or audio-frequency carrier by varying the amplitude of the modulating frequency.

Amplifier—an electronic circuit that provides gain.

Amplitude—the extent to which the voltage or current of an RF or audio signal varies compared to zero or a mean level.

Analog signal—an electronic signal that is represented by varying levels over a continuous range rather than in discrete steps.

Anomalistic period—a value for the period of an orbiting spacecraft determined by dividing the total number of minutes per day (1440) by the number of orbits per day.

Apogee—the high point (greatest distance from the surface) or a satellite orbit. (see Eccentricity and Perigee).

APT—*Automatic Picture Transmission*. An acronym describing the automatic transmission of images by polar-orbiting spacecraft.

Ascending pass—a pass originating to the south of a ground station and ending to the north of it. Ascending passes produce inverted images on display systems that don't account for the directional difference between a north-south and south-north pass.

Assembler—a computer program using mnemonic commands that permits the programmer to create, in one or more steps, a machine-language program for a specific microprocessor.

Azimuth—the compass direction (usually expressed in true degrees) required to point a ground station antenna at a distant spacecraft.

Az-el—an antenna mount that provides independent control of the azimuth and elevation of an antenna.

Bandpass filter—see Filter.

Bandwidth—the range of frequencies passed by a filter circuit. Since filters are never perfect, bandwidth is usually expressed as the total frequency range producing a 3-dB decrease in signal level.

BASIC—*Beginners All-purpose Symbolic Instruction Code*. The most widespread of the high-level programming languages for microcomputers that uses English-like statements and commands to create a computer program. There are generally two kinds of BASIC operating systems. In the most common form, interpreted BASIC, the BASIC statements or commands are converted to machine-language instructions at the time the statement or command is executed. This is the slowest form of BASIC. In *compiled* BASIC systems (such as QuickBASIC), a separate *compiler* program converts the BASIC commands and statements into machine-language code that represents the actual computer program. Compiled BASIC programs are notably faster in execution than interpreted programs because run-time conversion of the statements and commands is unneccessary.

Beamwidth—the width of an antenna pattern, usually expressed in degrees, that results in a 3-dB decrease in signal level at the beamwidth pattern limits.

Bearing—the combination of elevation and azimuth required to point an antenna at a specific spacecraft at any given moment.

Bidirectional—a data port through which data can travel both to and from a specific device.

Binary—a system of expressing numerical values to the base 2.

Bit—the smallest unit of data in a computer system. Any specific data bit can either be high (1) or low (0).

CDA—*Control and Data Acquisition*; acronym for the ground control stations for the TIROS/NOAA polar orbiters.

Chip—vernacular term for an integrated circuit.

COSMOS—a generic name applied to a wide range of Soviet research satellites. Some COSMOS spacecraft transmit very-high-quality 240-LPM meteorological images.

CRT—*Cathode Ray Tube*; a television-like image display device.

Crystal—one or more thin plates of quartz, mounted in a hermetically sealed enclosure. Crystals are used to control the frequency of oscillator circuits and are used as passive filter elements in receiver IF circuits.

Decay—the gradual decrease in orbital period and spacecraft orbital altitude caused by residual atmospheric drag.

Descending pass—a pass that originates to the north of a ground station and terminates to the south of it.

Downconverter—a series of modules, usually consisting of one or more RF preamplifiers, a local oscillator, mixer, and IF amplifier designed to convert S-band (1691-MHz) signals to the 136 to 138-MHz VHF satellite band.

Eccentricity—the ratio of apogee and perigee. A circular orbit has an eccentricity of 1.

Elevation—the bearing in degrees, relative to the horizon, of the principal lobe of an antenna. At an elevation of 0° the antenna is aligned with the horizon. At an elevation of 90° it is pointing straight up.

ERP—*Effective Radiated Power*. The effective power output of a transmitting system (usually expressed in watts or milliwatts (thousandths of a watt), taking into consideration the power output of the transmitter, losses in the transmission line between the transmitter and the antenna, and the antenna gain and pattern characteristics.

ESRO—*European Space Research Organization*. A consortium of European nations operating specific space-research projects such as the METEOSAT geostationary satellite system.

Facsimile—a system for reproducing a satellite (or other) image directly onto photographic or other specialized paper media or a computer display.

Fax—another word for facsimile.

Feed horn—one of the simplest systems for intercepting the RF energy reflected by a parabolic antenna and transferring that energy to a transmission line.

Filter—an electronic circuit composed of either passive (crystals, ceramic filters, coils and capacitors) or active devices (usually operational amplifiers) designed to pass a certain range of frequencies while providing some measure of reduction in signals at other frequencies. *Band-pass* filters (such as the crystal and ceramic filters in receiver IF circuits), or the active filter at the input of the video circuit of Chapter 5, are designed to pass a specific range of frequencies (defined by the design center frequency and bandwidth of the filter), while rejecting signals above and below those frequencies. *Low-pass* filters, such as those in the post-detector video circuits, are designed to pass signals below a specific frequency, rejecting signals above that frequency. *High-pass* filters reject signals below a specific frequency.

Focal length—the distance from the center of the surface of a parabolic antenna to the point at which the RF energy is brought to a focus (the focal point).

Geometric distortion—foreshortening of a view toward the horizon resulting from the wide field of view of the image-scanning radiometers. This distortion was obvious on the ITOS/NOAA spacecraft (NOAA 2-5), but is essentially removed by on-board computer processing of the HRPT data stream in the TIROS/NOAA (NOAA 6-11) spacecraft. (Only NOAA-9, -10 and -11 are presently active.)

Geostationary orbit—an orbit (approximately 22,000 miles high) the plane of which lies on the equator. Such an orbit has a period of 24 hours (1440 minutes). If the direction of movement of the spacecraft in this orbit is the same as the direction of the rotation of the earth below, the spacecraft subpoint on the equator will not change.

GIF—*Graphical Interchange Format*. A set of graphics standards developed by CompuServe that permits grayscale images to be displayed on a wide range of computers with different graphics standards.

GMS—*Geostationary Meteorological Satellite*. A spacecraft, similar to GOES, which is operated by the Japanese government over the western Pacific.

GMT—an obsolete term; see UTC.

GOES—*Geostationary Operational Environmental Satellite*. The geostationary meteorological satellite system, consisting nominally of three GOES spacecraft, operated by the US Government.

Ground track—projection on a map of successive spacecraft sub-point positions (see Track).

Horn—see Feed horn.

HRPT—*High Resolution Picture Transmission*; acronym for the high-resolution imaging system employed by the TIROS/NOAA spacecraft. Medium-resolution APT data are derived by sampling this data stream using the on-board computer.

Inclination—the angle of the orbital plane of a spacecraft relative to the earth's equatorial plane.

Infrared—(IR) electromagnetic energy with wavelengths somewhat longer than that of the visible-light spectrum. IR energy is essentially various wavelength of heat radiation lying beyond the red end of the spectrum.

Injection—the process of insertion of a satellite into orbit at a precise angle and speed to achieve the desired orbital geometry.

Interlaced—a system by which a complete image is built up on a CRT by scanning one field consisting of even-numbered lines followed by a second field of odd-numbered lines. Such a system is used in broadcast TV to maximize display resolution while minimizing flicker. Each field requires 1/60th of a second for display; the entire interlaced picture (a *frame* consisting of two fields) is produced in 1/30th of a second.

Intermediate Frequency (IF)—a frequency to which an RF signal is converted (usually lower than the signal frequency in most simple receivers) where it is more convenient to obtain the required signal gain while filtering the signal to obtain a specified bandwidth.

IR—see infrared.

Keplerian elements—a moderately complex set of numerical values that completely describes the nature of the orbit of a spacecraft. Keplerian elements for most satellites are routinely compiled by NASA and are available through a number of satellite bulletin board systems. Sophisticated tracking programs use such elements to gain the required precision for accurate tracking.

LPM—*L*ines *P*er *M*inute; defines the horizontal scanning rate of the spacecraft imaging system that must be duplicated for proper image display on the ground-station equipment.

Meteor—the primary polar-orbiting spacecraft series operated by the Soviets. Meteor spacecraft have slightly different orbital characteristics, depending upon the series, but all provide 120-LPM visible-light imagery. The latest spacecraft of this series have been testing IR imaging systems.

METEOSAT—the geostationary spacecraft operated by the European Space Research Organization.

Mosaic—A composite image created by combining individual display segments to produce an image with greater geographical coverage. The cover photo is a mosaic consisting of four segments: the northern and southern segments of an overhead and western pass.

Multi-spectral—an imaging system with individual sensors that respond to different portions of the spectrum. Current TIROS/NOAA spacecraft have a 5-channel instrument with one visible-light sensor and four IR sensors.

NOAA—*N*ational *O*ceanic and *A*tmospheric *A*dministration; spacecraft: the designation of a TIROS spacecraft after launch; agency: NOAA is the agency of the US Department of Commerce responsible for US governmental meteorological services.

Nodal period—the period, in minutes, between successive north-bound equatorial crossings of a spacecraft.

Noise—essentially random RF energy originating in signal processing circuits or from external man-made or astronomical sources.

Non-interlaced—a display system where the final image is created by a single scan of the CRT display tube.

Path loss—the difference between the effective radiated power of an RF source and the power available at the receiver site. Path loss is related to the distance between the source and receiver and the operating frequency.

Perigee—the low point in a satellite orbit. (See Apogee and Eccentricity).

Period—a general term for the time required for one orbit of the earth (see Anomalistic Period and Nodal Period).

Phasing—the processing of getting the display system in step with the image transmitter so that the start of image lines from the transmitter corresponds to the start of a line on the display. An out-of-phase image has the start of one image line somewhere in the main image area rather than at the edge (usually left) of the image.

PEL—*P*icture *El*ement (see pixel).

Pixel—the individual component of an image scan line defined by a single video sample in a digital image system. The greater the number of pixels in a line, the greater the display spatial resolution.

Polar mount—a mounting system for geostationary antennas where the main axis of the system is oriented parallel to the earth's polar axis. This permits an antenna to sweep the entire geostationary arc (Clarke Belt) using a single motorized drive, as opposed to individual control of azimuth and elevation.

Polar orbit—strictly speaking, an orbit whose plane intersects the poles of the earth. In practice, the term is loosely applied to the near-polar orbits of the TIROS/NOAA and Meteor/COSMOS spacecraft.

Probe—the RF pick-up element in an S-band feed horn.

Programmable scanner—a scanning receiver whose operating frequencies are controlled by a frequency synthesizer set by front-panel switches.

Quadrant (quads)—segments of the earth disk transmitted via WEFAX through the GOES geostationary spacecraft.

Rotator—a motorized system for remotely controlling the azimuth or elevation of an antenna system.

S-band—a relatively low-frequency microwave band including the 1691 and 1694.5-MHz frequencies used by geostationary weather satellites.

S/N—*Signal* to *Noise* ratio; the relationship of the power of the desired signal relative to total received power, including undesired noise. The higher the signal to noise ratio, the better the quality of the received signal.

Squelch—an electronic system for cutting off the audio output of a receiver when no signal is present.

Subcarrier—an audio tone (typically 2400 Hz) used to convey the amplitude-modulated video data.

Subpoint—The point on the earth's surface immediately below a spacecraft at any point in time.

Sun synchronous—an orbit in which the orbital plane precesses at slightly less than one degree per day, resulting in a situation in which the spacecraft passes overhead at essentially the same solar time throughout all seasons of the year.

Synchronization—the process of matching line scanning rate in the ground-station display equipment with that of the image transmitting system. Very slight synchronization errors result in tilted or skewed images, while larger errors render the image unrecognizable.

Synchronous motor—a motor designed to maintain an operating speed that tracks the frequency of the ac power signal. Such motors, controlled by a crystal or tuning-fork time standard, are used to maintain synchronization in electromechanical fax recorders.

Synthesizer—a complex of digital circuits that create or synthesize an RF or audio signal at some specific frequency.

TIROS—*T*elevision *I*nfra*R*ed *O*perational *S*atellite; the pre-launch or series designation for the current US polar-orbiting weather satellites. The acronym is the same one used for the very first operational weather satellites. (It is a bit of an anachronism in that none of the current spacecraft employ the television vidicon tubes of their predecessors.)

TNL—*T*hermal *N*oise *L*evel; the internal noise power level (usually expressed in dBmw [decibels per milliwatt]) of a receiving system. TNL is largely determined by the gain and noise figure characteristics of the early RF amplifiers in a receiving system. The TNL should be as low as possible because the desired signal must exceed TNL by 10-20 dB to achieve noise-free images.

Track—the process of adjusting the elevation and azimuth of an antenna to follow a spacecraft for the duration of a pass.

UTC—the correct term to use for time referenced to the prime meridian. Other commonly used terms are GMT (obsolete) or Z (Zulu) time. All spacecraft data are disseminated in terms of UTC time and date, but labeled GMT.

VAS—*V*ertical *A*tmospheric *S*ounder; the high-resolution, multi-spectral imaging system employed by the GOES spacecraft.

VISSR—*V*ery high-resolution *I*nfrared *S*pin *S*can *Ra*diometer; the GOES imaging instrument.

WEFAX—*W*eather facsimile; an acronym (*WE*ather *FA*X) for the medium-resolution imaging service, including pictures and charts, provided by the various geostationary weather satellites.

Parts and Equipment Suppliers

Unlike the situation a few years ago, there is now a comparatively large number of companies servicing the needs of weather-satellite experimenters. Throughout this book, I have noted specific manufacturers and products. Some were discussed at greater length than others. This doesn't neccessarily represent my assessment of the relative worth or value of the products—it simply reflects my familarity with the equipment line or product on the basis of my own experience. Because I cannot hope to sample everything out there—time and finances don't allow it—my experience cannot possibly be all-inclusive. Similarly, the listing that follows is not complete because since I cannot hope to be familiar with all of the possible vendors and products. Old products are upgraded or replaced, and new models and companies enter the field.

Computer bulletin boards, JESAUG, and reviews and advertisements in Amateur Radio journals such as *QST*, *73 Magazine*, and *CQ* (and their peer publications in other areas of the world), can help to keep you up to date on new product lines. By all means, get the manufacturer's literature and photographic reproductions of typical system output, if possible. Remember: Systems designed for HF facsimile use may *not* be compatible with the FM/AM format used by weather satellites. Also, some systems may produce reasonable output on weather charts, but may be deficient in handling gray-scale imagery. The weather-satellite images included in this book represent the kind of quality you can expect with a very modest investment in your station.

The following key will be used in the *Products* line following each supplier:

P—General electronics parts
S—Scan-converter hardware/software
T—Tracking software/hardware
SA—S-band antenna systems
SC—S-band converters
SP—S-band preamplifiers
VA—VHF antenna systems
VP—VHF preamplifiers
VR—VHF receivers
X—Crystals

SUPPLIER LISTING

A&A Engineering, 2521 W LaPalma #K, Anaheim, CA 92801, tel 714-952-2114.
Products: P, S.

Advanced Receiver Research, PO Box 1242, Burlington, CT 06013, tel 203-582-9409.
Products: VP.

Alpha Products, 242-B West Avenue, Darien, CT 06820, tel 203-656-1806; fax 203-656-0756.
Products: S.

Amateur Electronic Supply, 4828 Fond du Lac Avenue, Milwaukee, WI 53216, tel 800-362-0290.
Products: VR.

American Radio Relay League (ARRL), 225 Main St, Newington, CT 06111, tel 203-666-1541, fax 203-665-7531.
(A complete ROM listing—see Chapter 5—can be obtained for $3. Contact the Technical Department Secretary and request the *Weather Satellite Handbook* Metsat ROM listing.)

If you've purchased this book without the optional *Weather Satellite Handbook* Utilities Disk and want to order the disk, contact the ARRL Publication Sales Dept. The disk price is $10. Shipping and handling charges: $3.50 via UPS; $2.50 via US third-class mail in the US; foreign orders, $4.00 for surface mail.

AMSAT, PO Box 27, Washington, DC 20044,
tel 301-589-6062.
Products: T.

Applications Techniques, 10 Lomar Park Drive,
Pepperell, MA 01463, tel 508-433-5201.
Products: S (Pizazz Plus).

Cushcraft Corporation, 48 Perimeter Road,
Manchester, NH 03108, tel 603-627-7877; fax
603-627-1764.
Products: VA.

DARTCOM, Postbridge, Yelverton, Devon UK,
PL20 6SY.
Products: VR.

Down East Microwave, Box 2310, RR 1, Troy, ME
04987, tel 207-948-3741.
Products: SP.

David E. Schwittek, 1659 Waterford Rd, Walworth,
NY 14568, tel 315-986-2719.
Products: S.

Greg Ehrler, 105-53 87th Street, Ozone Park, NY
11417, tel (unpublished).
Products: SP.

Hamtronics Inc, 65C Moule Road, Hilton, NY 14468,
tel 716-392-9430; fax 716-392-9420.
Products: VP, VR (kits and wired and tested).

International Crystal Manufacturing, 10 N Lee,
Oklahoma City, OK 73102, tel 405-236-3741.
Products: X.

Jameco Electronics, 1355 Shoreway Road, Belmont,
CA 94002, tel 415-592-8097; fax 415-592-2503 and
415-595-2664.
Products: P.

JDR Microdevices, 1224 South Bascom Avenue, San
Jose, CA 95128, tel 800-538-5000; fax
408-559-0250.
Products: P.

L. L. Grace Communications Products, 41 Acadia
Dr, Voorhees, NJ 08043, tel 609-751-1018.
Products: T.

MetraByte Corporation, 440 Myles Standish Blvd,
Taunton, MA 02780, tel 508-880-3000; fax
508-880-0179; BBS 508-824-8659.
Products: S.

Metsat Products, Inc, 1257 Genmeadow Lane,
East Lansing, MI 48823, tel 517-332-7665.
Products: S, SA, SC, SP, VR.

Quorum Communications, Inc, PO Box 277,
Grapevine, TX 76051, tel 817-488-4861.
Products: S, SA, SP, SC, VP, VR.

Satellite Data Systems, PO Box 219, Cleveland, MN
56017, tel 507-931-4849.
Products: S, SA, SP, SC, VP, VR.

Software Systems Consulting, 150 Avenida Cabrillo,
Suite C, San Clemente, CA 92672,
tel 714-498-5784.
Products: S.

Spectrum International, PO Box 1084, Concord, MA
01742, tel 508-263-2145.
Products: SA, SC, SP, VA, VP.

Vanguard Labs, 196-23 Jamaica Avenue, Hollis, NY
11423, tel 718-468-2720 (Monday-Thursday,
8 AM-1 PM); BBS 718-740-3911.
Products: VA, VP, VR.

Wilmanco, 5350 Kazuko Court, Moore Park, CA
93021, tel 805-523-2390; fax 805-523-0065.
Products: SA, SC, SP.

Scan-Converter Parts List

The following parts list has been organized to make it easier to allocate parts when using the Metsat PC board set. The *Total Req'd* column represents the number of parts that are needed. These parts can be obtained from many sources. For those ordering via mail, the vendors in Appendix I (P product key) provide reliable service, good prices, and quality components. The remaining columns in the following list summarize the use of each part on the CPU and Display circuit boards and in the mainframe wiring.

Description	CPU	Display	Main-frame	Total Req'd
Diodes				
1N914	2	14	–	16
1N757 9.1-V, 0.5-W Zener	1	–	–	1
1N4004	1	–	1	2
1N4733 5-V, 1-W Zener	1	–	–	1
Transistors				
2N4401	–	2	–	2
TTL ICs				
74LS00	–	3	–	3
74LS138	1	–	–	1
74LS157	–	1	–	1
74LS221	–	1	–	1
74LS244	–	6	–	6
74LS374	–	1	–	1
74LS393	–	4	–	4
CMOS IC				
CD4020	–	1	–	1

Description	CPU	Display	Main-frame	Total Req'd
Linear ICs				
LM324	2	–	–	2
NE555V	1	–	–	1
NE567	–	1	–	1
7805 5-V regulator	1	–	1	2
Microprocessor/Memory ICs				
6809 CPU	1	–	–	1
6821 PIA	1	–	–	1
43256 RAM (see Notes)	1	1	–	2
Analog/Digital Converter IC				
ADC0804	1	–	–	1
IC Sockets				
8-pin DIP	1	1	–	2
14-pin DIP	2	7	–	9
16-pin DIP	3	4	–	7
20-pin DIP	1	7	–	8
28-pin DIP	2	1	–	3
40-pin DIP	2	–	–	2
Fixed Resistors, Metal-Film (1/4 W, 5%)				
10 ohm	–	1	–	1
150 ohm	1	1	–	2
220 ohm	–	1	–	1
470 ohm	–	4	–	4
510 ohm	–	1	–	1
620 ohm	–	1	–	1
680 ohm	–	1	–	1

Description	CPU	Display	Main-frame	Total Req'd
1 kilohm	1	3	–	4
1.2 kilohm	–	1	–	1
1.5 kilohm	–	1	–	1
2.2 kilohm	4	1	–	5
2.4 kilohm	–	1	–	1
4.7 kilohm	14	6	–	20
10 kilohm	12	3	–	15
15 kilohm	–	1	–	1
20 kilohm	3	–	–	3
120 kilohm	1	–	–	1

PC-Mount Potentiometer (¼ W, TO-3 pattern)

Description	CPU	Display	Main-frame	Total Req'd
5 kilohm	–	1	–	1

Panel-Mount Potentiometer (½ to 2 W)

Description	CPU	Display	Main-frame	Total Req'd
10 kilohm	–	–	1	–

Trimmer Capacitor (range not critical)

Description	CPU	Display	Main-frame	Total Req'd
3-20 pF trimmer	–	1	–	1

(Any value in this general range will suffice.)

Dipped Silver-Mica Capacitors (pF)

Description	CPU	Display	Main-frame	Total Req'd
10	–	1	–	1

(Values between 10 and 20 pF are satisfactory.)

22	2	–	–	2

(Values between 20 and 30 pF are satisfactory.)

150	1	–	–	1

(Values between 150 and 220 pF are satisfactory.)

Dipped Mylar Capacitors (µF, 50 WVDC)

Description	CPU	Display	Main-frame	Total Req'd
0.001	–	1	–	1
0.01	6	–	–	6

Description	CPU	Display	Main-frame	Total Req'd
0.1	1	–	–	1
0.22	1	–	–	1

Disc-Ceramic Capacitors (µF, 50 WVDC)

Description	CPU	Display	Main-frame	Total Req'd
0.1	13	6	2	21

Dipped Tantalum Electrolytic Capacitors (µF, 16 WVDC min)

Description	CPU	Display	Main-frame	Total Req'd
1 3	–	–	3	
2.2	–	1	–	1
4.7	–	1	–	1
10	5	3	–	8

Miscellaneous Components

Description	CPU	Display	Main-frame	Total Req'd
4.00-MHz crystal	1	1	–	2
4.194304-MHz crystal	–	1	–	1
LED	1	–	–	1
Phono Jacks	–	–	4	4
Normally-open push-button switch	–	–	4	4
5-position (min) rotary switch	–	–	1	1
16-conductor header cable with 16-pin male DIP plugs at each end	–	–	2	2
4-pin mike jack	–	–	1	1
4-pin mike plug	–	–	1	1

Notes:

The 43256 (Hitachi), M5M5256 (Mitsubishi) and CXK58256 (Sony) 32-kbyte static RAM chips have been successfully used on the display and microcontroller boards. All of these chips perform flawlessly, be they 150, 120, or 100 nS ICs. One chip that *has caused a problem* is the UM62256 (United Microelectronics Corp). The problem most often seen is random streaking of the display caused by random write errors. The displayed image is quite stable, but you can't get a clean write to the display memory. The problem isn't speed related, as far as I can tell. Until more data are available, I recommend that builders avoid using the UM62256 chip in the Metsat scan converter.

The 4-pin microphone jack and plug (available at Radio Shack stores [RS 274-002 and 274-001, respectively] and other suppliers) are used for the POWER connectors.

The two PC boards from Metsat Products (see Appendix I) fit neatly into an LMB (2946 E 111th Street, Los Angeles, CA 90023) M-E 7114 cabinet (4 × 11 × 7-inch [HWD]). Remove the internal chassis panel and mount the display board to the base of the cabinet. Mount the microcontroller board near the top of the cabinet on 1- × 1-inch extruded aluminum rails bolted to the front and rear panels of the cabinet.

WSH1700 BASIC Program Listing

```
10 KEY OFF
20 SCREEN 0
30 MODE$ = "None"
40 REM ===========================================================
50 REM MODULE 01 - define port values and initialize ports
60 REM ===========================================================
70 REM define ports
80 INPORT = 512
90 OUTPORT = 513
100 CONTROLPORT = 515
110 REM initialize ports
120 OUT CONTROLPORT, 144: REM set A for input and B for output
130 OUT OUTPORT, 255: REM set output to standby
140 REM ===========================================================
150 REM MODULE 02 - Main Menu
160 REM ===========================================================
170 REM main menu
180 GOSUB 2470
200 X$ = "** MAIN MENU **": GOSUB 2410
210 PRINT
220 PRINT TAB(5); "MANUAL MODES"; TAB(34); "AUTO MODES"; TAB(63); "DISPLAY CONTROL"
230 PRINT
240 PRINT TAB(5); "<A>PT"; TAB(34); "<W>EFAX"; TAB(63); "<D>isplay"
250 PRINT TAB(5); "<2>40 LPM"; TAB(34); "<V>is NOAA"; TAB(63); "i<N>vert"
260 PRINT TAB(5); "<1>20 LPM"; TAB(34); "<I>R NOAA"; TAB(63); "<C>omplement"
270 PRINT TAB(5); "<R>eset All Modes"; TAB(34); "<M>ETEOR"
280 PRINT
290 PRINT TAB(5); "MISC. DATA"; TAB(34); "SYSTEM"; TAB(63); "1700 IMAGE"
300 PRINT
310 PRINT TAB(5); "Mode:"; TAB(34); "<Q>uit to DOS"; TAB(63); "<L>oad Image"
320 PRINT TAB(5); "Date:"; TAB(63); "<S>ave Image"
330 PRINT TAB(5); "Time:"
340 PRINT
345 X$ = "Set <CAPS LOCK> key and type desired selection....": GOSUB 2410
350 LOCATE 18, 11
360 PRINT MODE$
370 LOCATE 19, 11
380 PRINT DATE$
390 LOCATE 20, 11
400 TYME$ = TIME$
410 PRINT TYME$
420 MENU$ = "main"
490 REM scan keyboard and evaluate responses
500 GOSUB 2100
510 IF Q$ = "Q" THEN OUT OUTPORT, 255: SYSTEM
```

```
520 IF Q$ = "R" THEN OUT OUTPORT, 255: MODE$ = "None": GOTO 170
530 IF Q$ = "W" THEN MODE = 6: GOTO 690: REM wefax
540 IF Q$ = "A" THEN MODE = 5: GOTO 690: REM apt
550 IF Q$ = "2" THEN MODE = 4: GOTO 690: REM 240 lpm
560 IF Q$ = "1" THEN MODE = 3: GOTO 690: REM 120 lpm
570 IF Q$ = "V" THEN MODE = 21: GOTO 690: REM auto noaa vis
580 IF Q$ = "I" THEN MODE = 22: GOTO 690: REM auto noaa ir
590 IF Q$ = "M" THEN MODE = 23: GOTO 690: REM auto meteor
600 IF Q$ = "D" THEN MODE = 2: GOTO 1170: REM display
610 IF Q$ = "N" THEN MODE = 1: GOTO 1170: REM invert
620 IF Q$ = "C" THEN MODE = 7: GOTO 1170: REM complement
630 IF Q$ = "L" THEN GOTO 1660: REM load image
640 IF Q$ = "S" THEN GOTO 1240: REM save image
650 GOTO 500
660 REM ============================================================
670 REM MODULE 03 - set 1700 operating mode
680 REM ============================================================
690 OUT OUTPORT, MODE: REM send mode to 1700
700 IF MODE = 6 THEN MODE$ = "WEFAX": REM set menu mode tokens
710 IF MODE = 5 THEN MODE$ = "APT"
720 IF MODE = 4 THEN MODE$ = "240 LPM"
730 IF MODE = 3 THEN MODE$ = "120 LPM"
740 IF MODE = 21 THEN MODE$ = "Vis NOAA"
750 IF MODE = 22 THEN MODE$ = "IR NOAA"
760 IF MODE = 23 THEN MODE$ = "METEOR"
770 REM route manual modes to phase
780 IF (MODE = 3) OR (MODE = 4) OR (MODE = 5) THEN GOTO 840
790 REM route auto modes to autophase delay
800 GOTO 1030
810 REM ============================================================
820 REM MODULE 04 - wait for phase inputs
830 REM ============================================================
840 GOSUB 2470
850 PRINT
860 X$ = "** PHASE MENU **": GOSUB 2410
870 PRINT : PRINT : PRINT
880 X$ = "Key <0> to initiate phasing or <Q> to exit....": GOSUB 2410
890 GOSUB 2100
900 IF Q$ = "Q" THEN OUT OUTPORT, 255: MODE$ = "None": GOTO 170
910 IF Q$ = "0" THEN GOTO 930
920 GOTO 890
930 OUT OUTPORT, 48: REM initiate phasing
940 GOSUB 2470
950 PRINT
960 X$ = "** PHASE MENU **": GOSUB 2410
970 PRINT : PRINT : PRINT
980 X$ = "Key <0> to stop phasing or <Q> to exit ....": GOSUB 2410
990 GOSUB 2100
1000 IF Q$ = "Q" THEN OUT OUTPORT, 255: MODE$ = "None": GOTO 170
1010 IF Q$ = "0" THEN OUT OUTPORT, 4: GOTO 170
1020 GOTO 990
1030 IF MODE$ = "WEFAX" THEN GOTO 170
1040 GOSUB 2470
1050 PRINT
1060 X$ = "** AUTO PHASE MENU **": GOSUB 2410
1070 PRINT : PRINT : PRINT
1080 X$ = "Waiting for auto-phase (key <Q> to abort)...": GOSUB 2410
1090 Q$ = INKEY$
```

```
1100 IF Q$ = "Q" THEN MODE$ = "None": GOTO 170
1110 V = INP(INPORT)
1120 IF V <> 0 THEN GOTO 170
1130 GOTO 1090
1140 REM ============================================================
1150 REM MODULE 05 - 1700 display control
1160 REM ============================================================
1170 OUT OUTPORT, MODE: REM send display code
1180 FOR N = 1 TO 500: NEXT N: REM delay
1190 OUT OUTPORT, 255: REM reset output port
1200 GOTO 170
1210 REM ============================================================
1220 REM MODULE 06 - get and save current 1700 image
1230 REM ============================================================
1240 GOSUB 2470
1250 PRINT
1260 X$ = "** SAVE CURRENT IMAGE **"
1270 PRINT : PRINT : PRINT
1280 GOSUB 2320
1290 IF Q$ = "Q" THEN GOTO 170
1300 PRINT : PRINT
1310 REM input and format file name of the save
1320 INPUT "Filename for image (8 char. max., no extension)"; FILE$: CLS
1330 IF LEN(FILE$) > 8 THEN GOTO 170
1340 FILE$ = FILE$ + ".wsh"
1350 GOSUB 2470
1360 PRINT
1370 X$ = "** SAVE CURRENT IMAGE **": GOSUB 2410
1380 PRINT : PRINT : PRINT
1390 X$ = "Transferring Image from 1700....": GOSUB 2410
1400 REM begin image transfer
1410 DEF SEG = &H6000: REM define segment for image data
1420 MEM = 0: REM zero pointer in segment
1430 V = 0: REM dummy video value
1440 OUT OUTPORT, 16: REM send request for image
1450 X = INP(INPORT): REM get data from 1700
1460 IF X = 255 THEN GOTO 1450: REM if standby, try again
1470 IF X = 16 THEN GOTO 1450: REM if echo request, try again
1480 FOR MEM = 0 TO 65535!: REM set up loop
1490 X = INP(INPORT): REM get 1700 data
1500 IF X = 255 THEN GOTO 1580: REM if done, exit
1510 IF X = V THEN GOTO 1480: REM if video data has not changed, look again
1530 V = INP(INPORT): REM re-read data
1540 POKE MEM, V: REM store byte in ram
1550 OUT OUTPORT, V: REM echo byte back to 1700
1560 NEXT MEM: REM loop until all ram bytes have been sampled
1580 OUT OUTPORT, 255: REM return output to standby
1590 BSAVE FILE$, 0, &HFFFF: REM save the segment as a binary file
1610 DEF SEG : REM reset segment
1620 GOTO 170
1630 REM ============================================================
1640 REM MODULE 07 - load and image from disk and pass to 1700
1650 REM ============================================================
1660 GOSUB 2470
1670 PRINT
1680 X$ = "** IMAGE LOAD ROUTINE **": GOSUB 2410
1690 PRINT : PRINT : PRINT
1700 GOSUB 2320
```

```
1710 IF Q$ = "Q" THEN GOTO 170
1720 REM display available file (those with .wsh extension)
1730 CLS
1740 X$ = "** FILES ON CURRENT DISK **": GOSUB 2410
1750 PRINT
1760 FILES "*.wsh"
1770 PRINT
1780 INPUT "Input desired file (no extension) or <Q> to exit..."; FILE$: CLS
1790 IF LEN(FILES$) > 8 THEN GOTO 1660
1800 IF (FILE$ = "q") OR (FILE$ = "Q") THEN GOTO 170
1810 FILE$ = FILE$ + ".wsh"
1820 REM load the image file to the video segment
1830 DEF SEG = &H6000: REM set video segment
1840 BLOAD FILE$, 0: REM load binary file to ram
1850 REM begin image transfer
1860 GOSUB 2470
1870 PRINT
1880 X$ = "** IMAGE LOAD **": GOSUB 2410
1890 PRINT : PRINT : PRINT
1900 X$ = "Sending " + FILE$ + " to 1700...": GOSUB 2410
1910 OUT OUTPORT, 32: REM send transfer request
1920 X = INP(INPORT): REM get 1700 data
1930 IF X <> 32 THEN GOTO 1920: REM if request not echoed, then look again
1950 FOR MEM = 0 TO 65535!: REM loop from start to end of the video segment
1970 V = PEEK(MEM): REM get byte from ram
1980 OUT OUTPORT, V: REM send to 1700
2000 X = INP(INPORT): REM sample 1700 output
2020 IF X = 255 THEN GOTO 2060: REM if end of image, exit
2040 IF X <> V THEN GOTO 2000: REM if last byte not echoed, look again
2050 NEXT MEM
2060 DEF SEG : REM reset segment
2070 OUT OUTPORT, 255: REM set output to standby
2080 GOTO 170
2090 REM =========================================================
2100 REM MODULE 08 - keyboard input routine
2110 REM =========================================================
2120 REM if routine was not called from the main menu, bypass the
2130 REM time update and mode check functions
2140 IF MENU$ <> "main" THEN GOTO 2250
2150 IF TYME$ = TIME$ THEN GOTO 2190
2160 TYME$ = TIME$
2170 LOCATE 20, 11: PRINT TYME$
2190 REM if no current image, bypass
2200 IF MODE$ = "None" THEN GOTO 2250
2210 REM if current mode is WEFAX, bypass mode status check
2220 IF MODE$ = "WEFAX" THEN GOTO 2250
2230 X = INP(INPORT): REM get 1700 port data
2240 IF X = 255 THEN OUT OUTPORT, 255: MODE$ = "None": GOTO 170
2250 Q$ = INKEY$: IF Q$ = "" THEN GOTO 2100
2270 MENU$ = "none"
2280 RETURN
2290 REM =========================================================
2300 REM MODULE 09 - disk insert prompt
2310 REM =========================================================
2320 X$ = "Insert a file disk and key <R> when ready": GOSUB 2410
2330 X$ = "or <Q> to exit....": GOSUB 2410
2340 GOSUB 2100
2350 IF Q$ = "Q" THEN RETURN
```

```
2360 IF Q$ = "R" THEN RETURN
2370 GOTO 2340
2380 REM ===========================================================
2390 REM MODULE 10 - center printed text routine
2400 REM ===========================================================
2410 X = INT(LEN(X$) / 2): Y = 40 - X
2420 PRINT TAB(Y); X$
2430 RETURN
2440 REM ===========================================================
2450 REM MODULE 11 - program banner routine
2460 REM ===========================================================
2470 CLS
2475 X$ = "************************": GOSUB 2410
2480 X$ = "*    WSH1700  Program   *": GOSUB 2410
2500 X$ = "*         (C)1990        *": GOSUB 2410
2520 X$ = "* Dr. Ralph E. Taggart *": GOSUB 2410
2525 X$ = "************************": GOSUB 2410
2530 PRINT
2540 RETURN
2550 END
```

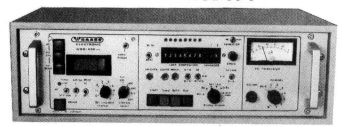

BRITISH AEROSPACE DARTCOM SYSTEM

- The world's fastest selling desk-top weather satellite image acquisition and display system.
- Receives all Met satellites, geostationary and polar orbiting.
- Expansion capability to accept all future APT Met satellites.
- Full spatial resolution without averaging.
- State of art reception, image display and processing.
- User friendly – menu driven under mouse control.
- Landsat and Spot images displayed and processed from floppy disk.
- Complete system for less than $10,000 (£6,000 sterling).

BRITISH AEROSPACE
SPACE SYSTEMS

Marketing Manager BAe Dartcom Systems
FPC 333 PO Box 5 Filton Bristol BS12 7QW
Telephone 0272 366379 Telefax 0272 366819 Telex 0272 449452

The Model 1700
A Better Alternative

The **Model 1700** scan converter from Metsat Products is the commercial version of the project Dr. Taggart designed as the center-piece of this new edition of his ***Weather Satellite Handbook***. Most of the images used to illustrate this edition were displayed using the **1700** so there is hardly any need to emphasize the image quality you can expect with this system. What distinguishes the **1700** from many other fine display systems on the market today is its unparalleled versatility. The **1700** scan converter can be operated in three distinctly different modes:

- As a **stand**-**alone** system, the **1700** will display **all** operational satellite image products on an inexpensive composite TV monitor. In this mode the **1700** will operate directly from the output of your receiver or from audio tape recordings.
- By interfacing viirtually **any** personal computer to the **1700**, the computer can be used to store images to disk. You need not have high resolution graphics capability since the **1700** itself takes care of all the basic display functions.

- If your computer has high resolution graphics capability, using it with the **1700** gives you state-of-the-art image display capabilities and almost unlimited opportunities for image processing, unattended operation, and a host of other options..

Unlike display systems that are linked to a specific bus, you need not worry about your investment becoming obsolete with new advances in computer and display technology. With the **1700**, your options for the future are unlimited. What's more, should your computer go down or otherwise be unavailable, your **1700** is still capable of providing basic image display, making it ideal for portable or emergency service as needed.

We can supply basic **printed circuit boards** and software **ROM**s for the home-builder or complete **wired and tested** systems. In addition, we provide complete software support for PC, XT, AT, and PS/2 compatible computer systems and are willing to work with software developers for any computer.

Call or write today for more information on the **Model 1700 - a better display solution for today and tomorrow**!

METSAT products, INC.

1257 GLENMEADOW LANE
EAST LANSING, MICH. 48823
PHONE: (517) 332-7665

Q U O R U M

Quorum Communications offers a complete line of components and systems for the reception of weather images for use by amateur experimenters, educators and professionals. No matter what your needs are, call us for our latest offerings.

On the following pages are data sheets describing two of our standard products, the IBM PC Wefax adapter and our SDC-1691B GOES downconverter. Also, we have a brief description of some of our other offerings below.

Unlike most other system providers, Quorum designs and manufactures ALL of our electronic components in house which allows us to provide the user with the highest quality and lowest cost available. If you have specific requirements, please give us the oportunity to quote your system.

Other Quorum Offerings

Receivers
Quorum offers stand alone and software programable receivers in various models capable of amateur and professional performance. Call for our latest catalog.

Down Converters
Quorum offers high performance, low noise down converters suitable for the reception of GOES Wefax, GMS Wefax, Meteosat and NOAA HRPT images. Weather proof and mast mount models are available.

Preamps
Various 137 and 1700 MHz preamp models are available for use with GOES, HRPT, Meteosat and APT systems.

Feeds and Antenna
APT antennas, TVRO dishes and 1700 MHz linear and circular feeds are available from Quorum.

Experimenters HRPT System
Readily available personal computers finally have the required computing power and storage requirements for high resolution digital HRPT images. Be at the leading edge of this exciting technology with Quorums HRPT components.

Call Quorum at (817) 488-4861 for our latest catalog or demo software.

QUORUM

SDC-1691B & SDC-1691BWP
1691 MHz Low Noise
Satellite Down Converter
June 1990

Features

- **Low Noise Figure**
  1.0 dB Typical
- **No external preamp required**
- **High Conversion Gain**
  33 dB Typical
- **Oven Stabilized Local Oscillator**
- **Compact Size**
- **Single Supply**
  + 12 to + 14 Volt Input
- **Temperature Stable**
- **Weather proof option available**
- **Options for HRPT and Meteosat**

Description and Usage

Reception of quality weather images from the USA GOES or European Meteosat geosynchronous satellites requires attention to detail particularly with respect to the signal reception system. The most important considerations for quality reception are antenna gain and downconverter performance. When used with a proper antenna, the SDC-1691B can easily provide images with 40 to 50dB signal to noise.

The most important feature of the SDC-1691B is it's excellent noise figure. When mounted at the antenna in a weatherproof enclosure, the 1 dB typical noise figure of the SDC-1691B provides state of the art performance. Because of the low noise figure, an external preamp is NOT necessary. This lowers the total system cost while providing increased reliability due to reduced component count. Along with a low noise figure, the SDC-1691B provides the necessary signal gain to overcome the loss of the IF feedline without affecting the receive system noise figure.

The second most important feature of the SDC-1691B is the use of an ovenized, temperature compensated crystal oscillator. The Wefax signal is very narrow, having a bandwidth of about 30KHz. Conventional downconverters drift due to temperature changes and require the user to continually re-tune the receiver to find the wefax signal. The SDC-1691B however, maintains it's output frequency within + - 2KHz from -20 to + 50 degrees C.

System performance is enhanced when the downconverter is mounted as close as possible to the antenna. The SDC-1691B can be upgraded to a weatherproof version with the SDC-WPA option, or it may be ordered from the factory as model SDC-1691BWP. The weatherproof option also allows mast or pole mounting.

Specifications

Noise Figure	< 1.2 dB, 1 dB typical
Conversion Gain	30 dB min, 33 dB typical
LO Frequency	1553.500 MHz
Image Rejection	20 dB
Output Frequency	137.500 MHz for 1691.0 MHz input
DC Power Input	+ 12 to + 14 V @ < 500 ma
Size	4.75" x 6" x 1" aluminum case
Input Connector	type 'N' female
Output Connector	BNC

Q U O R U M

WEFAX PC Adapter
Integrated Weather Fax Display Adapter
for the IBM PC / XT / AT and PS/2 model 30
June 1990

Features: (Version 2.50)

- Display of HF Fax, NOAA APT, GOES WEFAX, Goestap, Meteosat and Russian Meteor image formats
- Automatic aspect ratio correction
- Automatic Level Control for perfect image contrast
- Integrated Receiver controls for software selection of all operating parameters
- Tape recorder inputs and outputs with clock channel for perfect sync
- IBM EGA and VGA compatibility for display of up to 640 by 480 pixels in up to 16 levels of grey or colors
- Hot Key configurations can be user defined for instant switching between image modes and frequencies
- User configurable palettes for selection of 16 out of up to 262,000 colors
- User friendly menu driven interface
- Built in audio amplifier and keyboard volume control
- Automatic file save
- Automatic Image Capture
- Automatic synchronization, even in NOAA APT and Meteor modes
- Delayed operation via user configurable timer
- On screen signal strength when used with SL-137 receiver
- Sample Clock can be phase locked to the signal carrier for elimination of doppler effects on APT
- Images can be saved in GIF format
- Images can be printed to Epson 9 and 24 pin and Hewlett Packard PCL printers

Wefax PC Adapter Interface Card

Specifications (Version 2.50)

PC Bus Compatibility	IBM PC/XT/AT and compatibles
I/O Address	jumper selectable
Interrupt	jumper selectable 2 to 7
Card Size	10.25 inches long
Display Types Supported	IBM EGA and VGA
Memory Requirements	512KB min 640KB recommended
DOS Requirements	version 2.1 or later
Disk Requirements	hard disk highly recommended
Inputs	HF audio, APT and WEFAX audio, auxiliary audio, auxiliary clock and signal level
Outputs	Receiver control and data, signal level, speaker audio, tape audio, tape clock and receiver audio
Image resolution	Software version dependent 600 X 800 X 16 (Version 2.50)
Image file size	Image resolution dependent 240 KB (Version 2.50)

Quorum Communications, Inc. 1020 S. Main Street Suite A Grapevine, TX 76051 (817) 488-4861 FAX (817) 481-8983

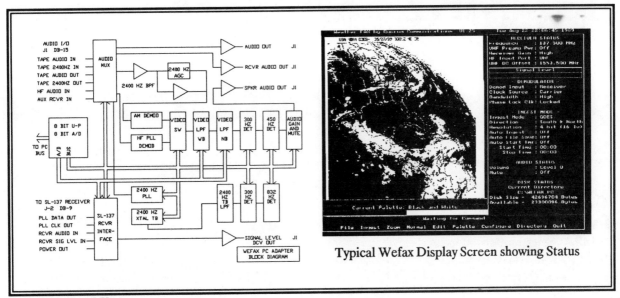

Typical Wefax Display Screen showing Status

Description and Usage

The Quorum Communications Wefax PC Adapter is an intelligent scan converter for HF Fax, NOAA APT, GOES, Goestap, Meteosat and Russian Meteor weather image formats. The Wefax PC Adapter contains an on board microcomputer with a built in 8 bit analog to digital converter, along with analog circuitry to select various audio input sources, control a slave SL-137 receiver and synchronize to and demodulate all current forms of weather facsimile. The high level of feature integration combined with the on board firmware and PC software provide a flexible platform for current and future performance.

Although the Wefax PC Adapter with its integrated microcomputer performs most of the work of signal reception, demodulation and synchronization, the user interface is a function of the PC software program provided with the adapter card. Unlike competing products, the user doesn't need to experiment with the correct number of clocks per pixel, rely on the accuracy of his PC oscillator or create image files that can only be used on the system they were captured on. The Quorum adapter and software take care of all the details. Simply identify the signal type, and the image is automatically captured and can be saved as a raw data file that can be used by anyone.

The current PC software revision (2.50) provides a rich set of user functions in an easy to learn and menu driven, fully graphical interface. It contains all the required features to capture, save to a file and read from a file in addition to quite a number of unique and useful features.

One of the most important features needed for reception of weather images is unattended operation. One necessary requirement for unattended operation is the ability to automatically detect an incoming image, synchronize to it and save it to disk for later viewing. All of these features are present in the current software and can be easily set up with user configurable hot keys.

When receiving APT images, doppler effects cause the image to bow and distort. The Wefax PC adapter allows the user to select a sample clock that is phase locked to the satellite signal carrier. When receiving NOAA APT images, the phase lock clock eliminates the doppler effects so that a geometrically correct image can be received. This feature can not be implemented with scan converters that rely on the PC clock for timing.

Most competing products require the user to change cables and/or receiver frequencies to switch between the various signal sources. The Quorum Wefax PC Adapter provides software selection of signal sources, and when used with our SL-137 receiver, signal frequencies. Typically, the user would configure the adapter and software operation for a particular signal type and then save the configuration to a hot key. Recalling the correct setup is the a simple key press. Individual configurations can be set up for all current satellites.

Any experienced weather image user can attest to the requirement for a large hard disk as image files quickly consume all available space. The Quorum PC software integrates file management so that images can be loaded, saved, renamed and deleted without returning to DOS. The Wefax PC Adapter allows you to record and playback signals on a stereo tape recorder, a feature competing systems don't offer.

Additional features, too numerous to mention here are also available. Call us for a demo disk.

137 MHZ SAT-TENNAS

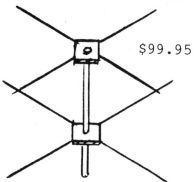

$99.95

$189.95

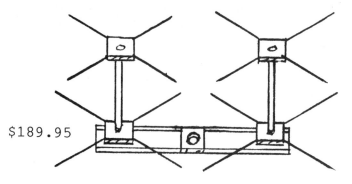

Type Wx-137

Type Wx-137A

CROSSED DIPOLES WITH REFLECTORS AND PHASE HARNESSES FOR CIRCULARLY
POLARIZED WEATHER SATELLITE RECEPTION. MADE OF HARDENED ALUMINUM,
FIBERGLASS, AND ABS PLASTICS. DEVELOPED FOR US/SOVIET WISP PROGRAM
AND IDEAL FOR TIROS AND METEOR WXSAT. INTERFACE TO 50 OHM COAX.
OTHER CONFIGURATIONS AND HARDWARE AVAILABLE. CALL/WRITE FOR INFO.

SHIPPING

EXTRA

 **RADIO ENGINEERS**
3941 Mt Brundage Ave.
San Diego, CA 92111
619-565-1319

CA AND FL.

ADD TAX

MULTIFAX SOFTWARE

This picture of Hurricane HUGO, was recorded at K2LAF from the NOAA 11 Satellite at 3 PM EDT, Thursday, September 21, 1989, using **MULTIFAX 4.0.** MULTIFAX SOFTWARE records, displays, manipulates, saves and prints weather facsimile pictures and charts from satellite and ground-based transmissions. It is designed for use with IBM PC's, XT's, AT's and clones.

FOUR PROGRAMS are available, each having features compatible with different equipment configurations:

MF2.3 for CGA, 320X200 pixels. Records, displays and prints in 2 or 4 colors from two pallets using an RGB monitor.

MF3.4 for EGA, 640X350 pixels. Records, displays and prints in 2, 4, 8 or 16 colors or shades from a group of 64 using EGA color or gray-scale monitor.

MF4.0 for VGA, 640X480 pixels. Records, displays and prints in 2, 4, 8 or 16 colors or shades from a group of 262,144 and provides gray-scale or color viewing with a VGA monitor.

MF5.0 for VGA, 640X480 pixels. Same as MF4.0, but for 2 and 16 colors or shades. For longer recordings it can automatically save to either a RAM disk in expanded memory (or hard disk). MF5.0 offers an enhanced syncing feature for Visual and IR APT WEFAX.

HIGHEST PICTURE DETAIL can be displayed because all programs sample and store the FAX signal into RAM 1280 times per FAX line. All detail may be viewed by zooming a selected area of the picture to fill the screen. Zooming increases detail as the viewing area is reduced, NOT by duplicating, averaging or dithering pixels. All programs also allow you to save all or any part of a picture or zoomed view to disk. All programs use the game port for data input, and can print on an IBM Graphics, IBM Proprinter or equivalent.

Call or write:
David E. Schwittek, NW2T
1659 Waterford Road, Walworth, NY 14568
315-986-2719

"PRECISION" AM/FM WEFAX DEMODULATOR

This new WEFAX Demod has been designed with quality and picture detail in mind. This Wefax Demod was created for use with NOAA and METEOR polar orbiting APT Facsimile, GOES geostationary facsimile, HF Weather Charts and gray-scale pictures, and HF News Service facsimile picture broadcasts. The switchable tuning display board makes this Precision AM/FM WEFAX Demod simple to use.

The PRECISION filters provide accurate, high bandwidth, response to facsimile signals, and give maximum detail. High quality components (1% resistors, 2% caps) insure long term stability, with no filter tuning. All adjustments are for levels only.

REQUIRES Facsimile Audio from receiver - speaker, earphone or line level, a Bipolar 12 volt power supply (+12v at 125 ma, -12 at 60 ma). A DVM, Scope and Calibrated Audio Generator are required for alignment.

AVAILABLE in a kit form, with all components, main and display PC boards and documentation. This kit requires a cabinet, power supply and connecting cables. A PC board set with complete documentation is available, as is a completly assembled and aligned kit (requires a cabinet, power supply and cables).

MULTIFAX - just $49
For any additional MULTIFAX program or future update, the price is just $20, postpaid in the U.S., Canada and Mexico. Add $3 for airmail elsewhere. No credit cards. Supplied on 2 - 5 1/4" disks with a full instruction manual. Sample picture disks are $2 in U.S., Canada and Mexico, $3 elsewhere. State CGA, EGA or VGA.

"PRECISION" AM/FM WEFAX DEMOD - just $129
in kit form, $29 for the PC board only kit with documentation, and $179 for the assembled kit. Kits are plus $5 shipping. COD extra.

MULTIFAX Software and **instruction books** are written and copyrighted by Elmer W. Schwittek.

Timestep

Timestep Weather Systems supply equipment to the education market, the amateur market and the professional market. Thousand of systems are in English and American schools. Fisher Scientific in Chicago are our agents for education. Prestigious professional customers such as the United Kingdom Ministry of Defense and the United States Air Force use our systems to make crucial judgments.

The London Science Museum chose our equipment to go on permanent display on the basis that it was the first affordable high performance system available on the open market.

We are one of the very few companies that design and manufacture all of our systems from scratch. Dave Cawley G4IUG our Managing Director heads the RF and Antenna design. Peter Arnold our new Cambridge University Graduate writes our powerful and innovative software. Dave Cutts G4FAW generates PCB artwork using the latest CAD packages and runs the Photo Plotter for fast design work. In a nutshell we design the antennas, the RF, the logic, the software, the PCB's and manufacture in house. This way we are completely in control of quality.

We use network analyzers, synthesised signal generators, a spectrum analyzer and tracking generator, CAD, a photo plotter, a wave solder machine and a vapour phase cleaner to achieve the best in value for money and quality. Open any of our equipment and you will find the best quality components on double sided, plate through and solder resisted PCB's. You will not find any hand wound coils, cut tracks or bits stuck up in mid air. The life of this ARRL book is going to exceed the life of our current products. If you are reading this at the end of 1990 or later please send for our latest catalogue that will have more exciting products !

We are looking for agents in all countries. If you advertise regularly and do not currently supply weather satellite equipment, please get in contact with us. We can supply all systems and provide the backup you and your customers will expect.

VGASAT III

This is a package that will take audio from a weather satellite receiver and allow the display and manipulation on a PC. It needs no calibration to the host computer at all. Featuring completely automatic acquisition that is displayable in up to 800 pixels 600 lines and 256 colours from GOES, Meteosat, NOAA and Meteor satellites. This system stores the full transmitted resolution on disc so that even if you have a CGA card at the moment you can later look in full detail when you get a SVGA card. All image files will be useable or convertible in future systems, making sure that your software and hardware is easily upgradable in the future.

This half card has two separate receiver inputs, one for Polar and one for Geostationary and can work live or from tape (even your old tapes !). GOES, Meteosat and NOAA are synchronised from their own subcarrier and the Soviet Meteor's are synchronised by Software Digital Signal Processing (SDSP).

The full resolution is stored on disc and can be displayed down to pixel level in zoom format, with any colour card from CGA upwards to 800x600x256 SVGA. All IBM register compatible cards are supported as well as Paradise, Orchid, Sota, Video Seven, Morse and other cards based on Paradise, Tseng and Trident (ZyMOS).

Animation from GOES and Meteosat is built in (EGA and VGA only). We call this "stand alone animation" as it will automatically take images, store them and display them continuously. Old images are automatically updated with new images. This feature is completely smooth and flicker free and will operate from 8 to 64 images depending on the memory available.

NOAA and Meteor are handled in a particularly innovative way. The software will look for the edge sync data and when it detects it's presence will automatically plot the image on screen. A user selectable delay for NOAA can be used that will eliminate horizon passes and start and stop acquisition so that your location is centred on the screen.

A fast HP LaserJet type printer routine is included. Taking just 2 minutes to process and allowing multiple copies to be made, this software will also allow a user comment line to be printed.

Full colouring and autoshade routines are built in to enhance images. Density slicing, contrast expansion and reduction are all built in. X-Y coordinate and pixel values aid temperature calibration. Images can be converted into GIFF format for easy exchange. Colorix VGA Paint can also be used to manipulate images.

The complete system including PC Card, all software and a manual is available for just $249.00 including delivery by Federal Express direct to your door.

Timestep Weather Systems
Wickhambrook Newmarket CB8 8QA England
Tel. +44 (0)440 820040   Fax. +44 (0)440 820281

Timestep

Goes/Meteosat 12 foot 55 element yagi. $149.00

Goes/Meteosat pre-amplifier. 28dB gain less than 1dB noise figure.   Fitted with female **N** connector to fix directly onto the feed with no co-ax cable. $149.00

Goes/Meteosat receiver. Not to be confused with convertors ! 1.7Ghz in and audio out. 2 channel and must be used with pre-amplifier above. Active AFC and threshold extension are just some of the may features this receiver has. $399.00

GOES Meteosat coffee tin type dish feed. Professionally manufactured and powder coat paint finished, this feed has both polarities available making it particulary useful for GOES and **Meteosat** or for combining for circular polarization for NOAA HRPT. $49.00

NOAA 2 channel receiver. 137.50Mhz and 137.62Mhz Full professional quality RF and IF filtering. Immune to cross and intermodulation $199.00

PROSCAN 8 channel scanning receiver. Performance as our NOAA 2 channel but with 8 channel crystal controlled scanning. Crystals are used to give the low noise and purity not normally associated with synthesized receivers. 137.30, 137.40, 137.50, 137.62, 137.80 and 137.85 MHz fitted are fitted as standard with 2 spare sockets. Intelligent squelch circuitry opens only on weather satellites. Squelch relay fitted to operate tape decks. Manual or automatic control with any number of channels being locked out. Also fitted with a bi-directional computer interface for fully automatic systems. Built to an impressively high engineering and visual standard, this receiver is available for $499.00

NOAA/Meteor turnstile antenna. Designed and built to give optimum omnidirectional coverage from 3 degrees to each horizon. Only available from agents in your country. $64.00

NOAA pre-amplifier. Designed to take the cable loss out of a system and not to revive old receivers and hence block new ones ! 13dB gain and a 4 pole RF filter will give you noise free images when used with our turnstile antenna $49.00

Coming soon (by the end of 1990)

VGASAT IV.   Like VGASAT II but enhanced with 3D for realistic three dimensional cloud detail and sideways looking Radar on Soviet satellites. New display drivers including, 1024x768 256 grey TIGA driver, 1024x768x256 Orchid (Tseng) driver, 1024x768x256 8514 driver. Estimated price to existing VGASAT III users $99.00

MEGA NOAA/Meteor. A software program to dump the entire pass from polar orbiters on to your hard drive in a massive 2,048K file (yes 2Mb !). Sections may be manipulated and temperatures directly read from the mouse pointer. 3D and the above 1024x768 drivers will be included. Estimated price to existing VGASAT III users $99.00

HRPT Receiver. A stand alone receiver with 1.7GHz input and data output. With signal strength meter and 2 channels fitted. Will need our pre-amplifier and dish feed together with a steerable 4 foot dish.

HRPT PC DATA Card. Designed by John DuBois and Ed Murashie, this card will take the data from our HRPT receiver and store directly onto your hard drive. A small scale image shows the area being scanned. Estimated price $399.00

HRPT Software. Similar in function and facilities to our VGASAT IV but making full use of the 24 bit colour capability of the Hercules Graphic Station Video Card (TIGA), giving stunning colour images by combining several sensors into one image. It will also work with TIGA to produce 1024x768 in 256 grey levels in a 1:1 zoom ratio to give all the data on screen from one sensor. Estimated price $199.00 (but call, it may be much less !)

In Great Britain we are able to supply complete systems including VGA cards, monitors and computers at very competitive prices. Call us for a quote.

New Ideas ??.   If you have any hardware or software that either needs tidying up or marketing World wide, then please write to us. We will redesign if necessary or just clean up the PCB layout. We will then give you the best exposure and marketing throughout the World.

WeatherWise Corporation
Specializing in High-Definition Weather Satellite Imagery

P.O. Box 73, Jarrettsville, MD 21084   (301) 577-9451

WeatherWise Corporation is devoted to offering total solutions to address all weather related system needs. These configurations include total turnkey systems geared towards educational institutions, plus, individual components for the do-it-yourselfer at affordable prices, with no compromise in quality.

- o **Receivers**
    Designed specifically for reception of ALL weather satellites frequencies, in both the VHF and UHF bands
- o **Antennas**
    Configurations include Turnstiles, Loop Yagis, Parabolic Dishes, etc.
- o **Preamplifiers**
    High Gain (> 20dB typ.), Very Low Noise (< 0.7db typ.) GaAs FET Preamps for both VHF and UHF bands
- o **Controllers, Processor Boards, and Test Equipment**
    Hardware/software packages to control and test the entire weather downlink station

- o **Detailed reference material** dealing with all Weather-related topics, including detailed satellite information; perfect for use in learning about what services are available, as well as teaching students fundamentals of Weather from a satellites perspective

- o **Bulletin Board Service** available by phone and amateur radio offering the most comprehensive source of data and information dealing with a multitude of weather related topics

*** Coming Soon: Weather Satellite Tutorial Seminars ***

Give us a call or drop us a line to get all the latest up-to-date information on our product line and new services available.

Weather Satellite Receivers, Down Converters and Antenna Systems for GOES, METEOSAT, NOAA, METEOR - WEFAX - APT - HRPT - VISSR/VAS

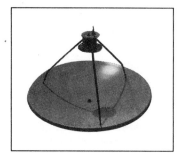

PARABOLIC
DISH ANTENNA WITH
WFDC-1691-137.5 FEED,
DOWN CONVERTER

MODEL M1S
SYNTHESIZED
WEATHER SATELLITE
RECEIVER

MICROSTRIP VHF
ANTENNA, RCP WITH
PRESELECTOR FILTER
AND GaAsFET PRE-AMP

FEATURES:

Receiver accepts VHF Antenna input as well as GOES/NOAA S-Band down converted signal. Outputs 2400Hz tone to an external scan converter. 21.4 MHz second IF has wideband output on rear panel to drive an external HRPT-VAS PLL demod. 45KHz bandwidth crystal filter and crystal discriminator for demodulation of WEFAX or APT. Two models - Synthesized Model M1S tunes any frequency 136-138MHz in 1KHz steps; Channelized Model M1 tunes up to six fixed frequencies.

Microstrip VHF Antenna has solid hemispherical coverage allowing up to 13 minutes of reception. Provides greater gain than crossed dipoles. Preselector filter and GaAsFet Pre-Amp, powered through coax, mount under antenna.

S-Band Down Converter has 20MHz bandwidth, utilizes thick film hybrids in machined structures. GaAsFET LNA is a stable, balanced amplifier design with NF of .75dB, Gain>35dB; high input return loss assures certainty of Noise Figure. Local Oscillator is crystal controlled, heater stabilized with <2ppm frequency stability from -20° to +40° C. 137.5 MHz IF amplifier has >30dB of gain. Entire assembly is powered through coax.

Call, write or FAX factory for latest catalog, technical data sheets, price list.
S-Band modules may be purchased separately.

5350 Kazuko Ct.
Moorpark, CA 93021

Phone: (805) 523-2390
Fax: (805) 523-0065

WEATHER SATELLITE
HANDBOOK

PROOF OF
PURCHASE

Please use this form to give us your comments on this book and what you'd like to see in future editions.

Name _____ Call sign _____

Address _____ Daytime Phone (___) _____

City _____ State/Province _____ ZIP/Postal Code _____

Edition   4 5 6 7 8 9 10 11 12

Printing 1 2 3 4 5 6 7 8 9 10 11 12

From_____

_____

_____

Editor, The Weather Satellite Handbook
American Radio Relay League
225 Main Street
Newington, CT 06111
USA

···························· please fold and tape ····························